Springer Proceedings in Earth and Environmental Sciences

Series Editor

Natalia S. Bezaeva, The Moscow Area, Russia

The series Springer Proceedings in Earth and Environmental Sciences publishes proceedings from scholarly meetings and workshops on all topics related to Environmental and Earth Sciences and related sciences. This series constitutes a comprehensive up-to-date source of reference on a field or subfield of relevance in Earth and Environmental Sciences. In addition to an overall evaluation of the interest, scientific quality, and timeliness of each proposal at the hands of the publisher, individual contributions are all refereed to the high quality standards of leading journals in the field. Thus, this series provides the research community with well-edited, authoritative reports on developments in the most exciting areas of environmental sciences, earth sciences and related fields.

More information about this series at http://www.springer.com/series/16067

Gennady Fedorov · Alexander Druzhinin ·
Elena Golubeva · Dmitry Subetto ·
Tadeusz Palmowski

Editors

Baltic Region—The Region of Cooperation

 Springer

Editors
Gennady Fedorov
Institute of Environmental Management,
Urban Development and Spatial Planning
Immanuel Kant Baltic Federal University
Kaliningrad, Russia

Elena Golubeva
Department of Geography
Moscow State University
Moscow, Russia

Tadeusz Palmowski
University of Gdańsk
Gdańsk, Poland

Alexander Druzhinin 🔟
North Caucasian Research Institute
Southern Federal University
Rostov-on-Don, Russia

Immanuel Kant Baltic Federal University
Kaliningrad, Russia

Dmitry Subetto 🔟
Department of Geography
The Herzen State Pedagogical University
of Russia
Saint Petersburg, Russia

Institute for Water and Environmental
Problems, Siberian Branch of the Russian
Academy of Sciences
Barnaul, Russia

Immanuel Kant Baltic Federal University
Kaliningrad, Russia

ISSN 2524-342X   ISSN 2524-3438 (electronic)
Springer Proceedings in Earth and Environmental Sciences
ISBN 978-3-030-14518-7   ISBN 978-3-030-14519-4 (eBook)
https://doi.org/10.1007/978-3-030-14519-4

This Springer imprint is published by the registered company Springer Nature Switzerland AG
The registered company address is: Gewerbestrasse 11, 6330 Cham, Switzerland

Preface

The Baltic Sea region (BSR) is traditionally considered a region of international cooperation developing at all levels—international, interregional and intermunicipal. Located in different countries, cities and towns, enterprises, social institutions and non-governmental organizations have established good neighbourly relations with their counterparts from the BSR. This fact has a positive impact on all spheres of public life: political, economic, social and cultural.

The BSR boasts various spatial forms of cross-border cooperation, including Euroregions, which facilitate the implementation of mutually beneficial projects launched by border areas of the neighbouring countries. The Council of the Baltic Sea States is an international body coordinating the activity of the BSR countries and regions. Researchers and academia from all countries of the region meet regularly to discuss topical issues of international cooperation at numerous conferences, round tables and seminars.

Kaliningrad has become a centre of international conferences, many of which have been hosted and organized by the Immanuel Kant Baltic Federal University (IKBFU). The university has close ties with many research centres located in the Baltic Sea region and researchers studying the socio-economic and political processes taking place in the BSR.

Saint Petersburg State University and the Immanuel Kant Baltic Federal University publish the "Baltic Region" journal in two languages—Russian and English. The journal is indexed in the Russian science citation system and the Web of Science and Scopus. The "Baltic Region" is on the list of the Higher Attestation Commission of the Ministry of Science and Higher Education of the Russian Federation, where doctoral and postdoctoral researchers publish their articles.

This collection of articles is based on the proceedings of the international conference "Baltic Region—the Region of Cooperation", held in Kaliningrad on 13–16 September 2017. The publication presents an overview of different research areas that geographers from Russia and Poland are interested in—economics, physical geography of the Baltic region, its spatial organization and the development of cross-border cooperation.

The articles show that despite the recent deterioration of political relations between Russia and its Baltic neighbours, cooperation in economic research, the social and cultural spheres and environmental protection continues.

Based on the research conducted by the Russian and Polish scientists in recent years, this collection of articles describes the current socio-economic development of the BSR countries as well as various forms of cross-border cooperation. The authors offer recommendations on the further development of the BSR. Special attention is paid to the study of interactions between Russia and the European Union in the Baltic Sea region. The agenda of this interstate cooperation includes environmental management and environmental protection, cross-border cooperation between enterprises, universities and people, the development of international tourism, expanding the contact function of the border and spatial and landscape planning.

A number of articles focus on the theoretical aspects of economic development and cross-border cooperation, international clustering, challenges of transition from the extensive to intensive economic development, etc.

Cross-border cooperation is an important factor in enhancing international competitiveness and ensuring the dynamic and sustainable development of neighbouring countries and regions. Therefore, the proposals substantiated in the book can be used for the development of strategies of economic development and intergovernmental projects under the umbrella of the Council of the Baltic Sea States.

The book is of interest to the regions, municipalities and institutions working on their strategies, as well as institutions and organizations of the Baltic region countries. Many of the questions discussed in the articles require further research and discussion.

Kaliningrad, Russia   Gennady Fedorov
Gdańsk, Poland   Tadeusz Palmowski
Moscow, Russia   Vyacheslav A. Shuper
Saint Petersburg/Barnaul/Kaliningrad, Russia   Dmitry Subetto
Rostov-on-Don/Kaliningrad, Russia   Alexander Druzhinin

Contents

Contemporary Applied Geographic Studies in the Baltic Sea Region: Public Geography

Geographic Research: Development of the Theory

Self-organization at the Trajectory Turn: Shift to Intensive Development

Vyacheslav A. Shuper

Abstract

In the development of the world economy, global integration and disintegration cycles alternate. Global economic growth can be viewed as hyperbolic by analogy with the population growth in S. P. Kapitsa's theory, meaning that its internal laws rather than external constraints determine it. S. P. Kapitsa formulated this concept as the principle of the demographic imperative. Its significant slowdown in the coming years is inevitable. There are two possible scenarios. One is pessimistic, "the Limits to the Growth"-like, the other is the shift from extensive to intensive development, which is similar to the ideas of the embattled Soviet philosopher Michael Petrov. There are several currently observed processes. First of all, it is the rapid intensification of industrial, agricultural and intangible production processes. Secondly, intangible consumption is largely replacing tangible one. The third prerequisite for social progress in the context of the significant slowdown in GDP growth rates is a reversal of the trend of progressive decrease in educational and intelligence level.

Keywords

Integration and disintegration cycles · Nonlinear dynamics · Hyperbolic growth · Intensive development

V. A. Shuper (✉)
Institute of Geography, Russian Academy of Sciences, Moscow, Russia
e-mail: vshuper@yandex.ru

V. A. Shuper
Department of Economic and Social Geography of Russia, Faculty of Geography,
Lomonosov Moscow State University, Moscow, Russia

G. Fedorov et al. (eds.), *Baltic Region—The Region of Cooperation*,
Springer Proceedings in Earth and Environmental Sciences,
https://doi.org/10.1007/978-3-030-14519-4_1

Introduction

"The current moment is characterized by the fact that it is static. And since it is not known when the static moment begins and when its end comes, I will first give my final remarks and then the opening ones" (V. V. Mayakovsky. The bathhouse). Comrade Pobedonosikov's speech before the launch of the time machine, i.e. before sending the expedition to the future, is of immediate relevance to the current situation, as it is characterised by the lack of understanding which direction the processes follow. There is a natural fear of change and no picture of a desirable future. There is Pobednosikov-like mentality of the shapers of our destinies and the rulers of our minds even in the most developed countries. The latter is explained by the fact that the folly of civilized people is no better than the folly of barbarians since there is only one kind of folly. Meanwhile, in the critical time, the familiar folly becomes an unaffordable extravagance.

Cycling of Integration and Disintegration Processes

The results of the US presidential election on November 8, 2016, as well the referendum on Brexit on June 23, 2016, and the constitutional referendum in Italy on December 4, 2016, were a clear demonstration of how deeply estranged the globalized elites are from the rest of the population. People of their countries have lost trust in them and no longer invest their hopes for a better life in them. This course of events is a natural result of the alternation of integration and disintegration cycles in the development of the world economy,[1] that was pieced together to form a coherent whole in the middle of the 19th century, when the electric telegraph made it possible to exchange information between stock exchanges in real time. Since in the middle of the XIX century England was producing half of the world's industrial output. It became the leader of the integration cycle that lasted until 1914. It went down in history as *Pax Britannica*. Its climax was *La Belle Époque*, which began in 1879 according to the French, and in the 1890s of the XIX century as believed by other countries. Sometimes this beautiful name stands for the fifteen-year period before World War I.[2] The first globalization was in many principal aspects deeper than the second one. Suffice it to say that visas were invented only after the World War I. In the world bristling with boundaries, the happy children of the first globalization cherished the sweet memories of their round-the-world trips made with a name card as the only proof of identity.

---

[1]Studies of the cyclical nature of integration processes were initiated in the late 1970s. in the Institute of Geography of the USSR Academy of Sciences B.N. Zimin (1929–1995) and continued LM. Sintzerov (Gorkin and Lipets 2003; Sintserov 2000).
[2]For France *Belle Époque*—the heyday of the Third Republic (in 1879, Marseillaise became a national anthem, in 1880, July 14 became a national holiday).

The disintegration of the world economy of 1914–1945 was not only the consequence of poor political decisions but also the result of objective socio-economic processes, which fall outside the scope of the text. We shall only note that the second global integration cycle, i.e. the ongoing globalization, began after the end of World War II and naturally went down in history as *Pax Americana*, as in 1945, the American economy accounted for a half the world's economy. It can be assumed that this period ended in 2008 with the peak of the current crisis. It is difficult to resist drawing an analogy and calling the "unipolar world" that existed between the collapse of the USSR and this crisis *Belle Époque 2.0*, as this was the culmination point of the second globalization. The classic signs of global disintegration have already appeared. They include the growth rate gap between world economy and world trade, as well as the rise of protectionism (they are obviously interrelated). We can assume that the third global integration cycle will begin in 15–20 years. However, there will be another leader. Presumably, it will be *Pax Sinensis et Indus*. It is the first time in 500 years that the global West has lost its leadership.

Current Changes from the Perspective of Synergetics

Apparently, the synergetic revolution is the last scientific revolution of the XX century, the last breakthrough of the era of the exponential growth of science. It absorbed its now inconceivable intensity of scientific creativity. The ideas of synergetics blossomed in socio-geographical studies (Shuper 2015). Important results were obtained by Yu. G. Lipets (1931–2006) and the Laboratory of Geography of World Development that he headed at the Institute of Geography of the Russian Academy of Sciences. The traditional for geography focus on spatial factor found its new materialisation in the notion of space as a source of development (Lipets and Pulyarkin 2001).

The works of S. P. Kapitsa (1928–2012) should be considered the most successful applications of the mathematical apparatus of nonlinear dynamics to geographical problems. The results obtained within the framework of his phenomenological theory of world population growth are widely known. Perhaps the most important result in terms of paradigm is that the growth of the Earth's population has never been constrained by external factors, i.e. conditions or resources, its pace has always been determined by the internal patterns of the process. "This fact allows formulating the principle of demographic imperative. It contrasts Malthus's population principle, according to which the resources determine the speed of population growth and its limit. The mathematics counterpart of the principle of the demographic imperative is the principle of subordination in Synergetics (Kapitsa 1999, p. 157)". These patterns and regularities expressed in nonlinear dynamics terms (so-called, hyperbolic growth) serve as the basis for the prediction of the Earth's population stabilization by the end of the XXIst century at the level of 12 bln. people, with 90% of this population (10.7 billion) to be reached

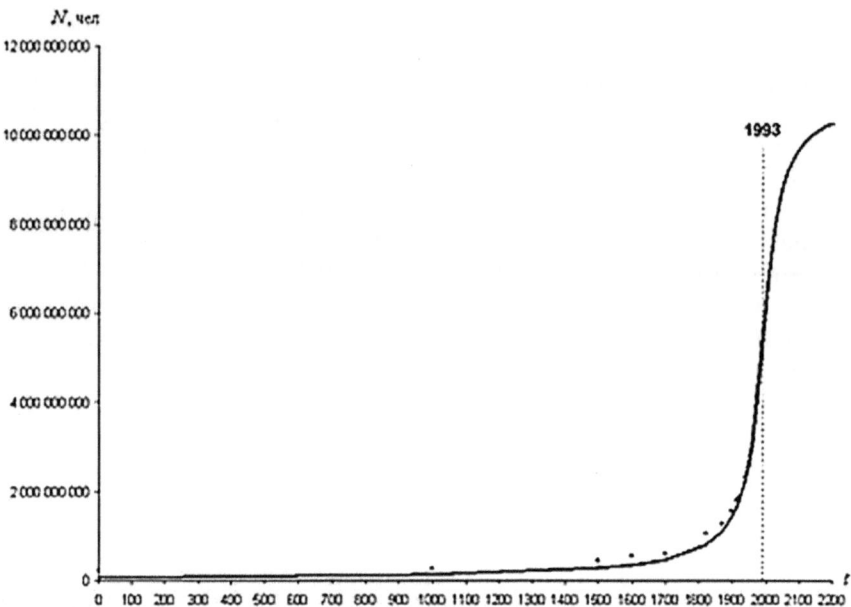

Fig. 1 Evolutionary model of the world population (Akaev and Akaeva 2011)

by mid-XXIst century. Figure 1 shows the results of calculations performed by A. A. Akaev and B. A. Akaeva using the model developed by S. P. Kapitsa.

S. P. Kapitsa continuously emphasised the fundamental relationship between demographic and historical processes. Population stabilisation and its subsequent gradual decline will be accompanied by profound changes in society to a certain extent caused by the population ageing. This author even raised the question of the inevitability of the return of some elements of the Middle Ages, which sadly seems less and less fantastic. It is reasonable to assume that the growth of the world economy is also hyperbolic, and can be described using the apparatus of nonlinear dynamics. "The Limits to the Growth" studies that flourished in the 1970s are identical to Malthusian fears and largely reflect them. The fact that the reality proves the fears of the exhaustion of natural resources to be unfounded, never-theless, does not imply the possibility of maintaining high rates of economic growth for an unlimited period.

The Slowdown in Economic Growth

The economic growth rate will certainly slow down, however, similarly to demo-graphic growth, this will happen not because of external constraints, but under the influence of its internal characteristics and patterns. The concept of sustainable development that does not stand any criticism appeared due to an exceptionally sharp economic slowdown in highly developed countries arising from the hollowing-out of the industry. It was necessary to convince the electorate that they were getting not the growing level of consumption but the better quality of life. "New normality" in China, which has become the powerhouse of the world economy, is a part of the same story. Even if the heroic efforts of the Party, the Government and all the great Chinese people to maintain the GDP growth rate of 6.5% required to build "xiaokang" moderately prosperous society for another 10 or 15 years are successful, the general trend will not change. The inflextion point is passed, and the world economy will gradually slow down.

It is common knowledge that the world is a product of high rates of economic growth to a much greater extent than we used to think. A. N. Pilyasov in his report at the Seminar on Regional Economics MSE MSU on April 6, 2017, reasonably assumed that the USSR could maintain its stability only at high rates of economic growth. Extremely slow economic growth dooms many institutions, including eminently respectable ones, to share the fate of the USSR. In many aspects, this will be underpinned by technological progress. Democracy loses ground due to the erosion of its social base, the middle class. This is exacerbated by the continuous improvement of technologies for manipulating the public consciousness. The extensive use of robots to meet sexual needs can become the last straw for the institution of the family. As in many other cases, virtual reality will inevitably materialise, since the robot's capabilities will by far exceed human ones. Moreover, robots will be most likely increasingly entrusted with child caring tasks. Hardly there will be no impact on the psychological formation of children. These are just a few examples that come to mind.

The Primacy of Human Potential

More optimistic scenarios are also possible. For example, crowdsourcing that confounded expectations will probably be replaced with the collective intelligence technologies (Slavin 2017). These technologies return to proven principles on a new technical level. One of them is personal responsibility, the absence of which inevitably corrupts a person. Anonymity will no longer exist, and evidence will be required. These technologies will prevent the endarkenment of society and facilitate the revival of public spirit, the basis of any normal democratic process. We must always remember that the future is uncertain, the trajectory at the bifurcation points can depend on the efforts of a small group of people or even an individual. E. Lorenz (1917–2008), one of the founders of chaos theory, called this phenomenon the "butterfly effect". Self-organization shows its effects in dissipative systems characterized by openness, non-equilibrium and non-linearity. Increasing non-equilibrium is not bound to lead to chaos, as it also facilitates the creation of order from chaos.

Without a doubt, humanity has now reached the bifurcation point, thus it can choose its development trajectory. It could be miserable mediocrity in keeping with "the Limits to the Growth", or the shift from extensive to intensive development. The concept was developed by the embattled Soviet philosopher M. K. Petrov (1923–1987), who managed to publish only a few articles during his lifetime, and nevertheless was highly regarded in the scientific community. The book that provided a comprehensive reasoning for the shift to intensive development was among his posthumously published works (Petrov 2004).

It was back in 70-80-s that M. Petrov put forward the ideas that exponential growth in the number of people employed in science could not last long, that further social development of society requires fundamental revision of the principles of education, science and other spheres of public life. It is not possible to expound the views of the insightful thinker on the need for a radically different education system, both secondary and higher, especially taking into account that some of his recommendations were bound to become outdated 30 years after his death. His book is not a manual for reformers, rather it is a proof of the necessity for *a new intellectual revolution,* the necessity recognised long before the extensive model evidently ran out of steam.

At the moment, there is no need to convince anyone of the catastrophic drop in the educational and intellectual levels, and the dramatic decline in the effectiveness of scientific research. In the 1970s, these trends have just been developing; nevertheless, they caused major concern among those who looked much farther ahead than most people did. It has become obvious that only a drastic reform in education and consequently in science can improve the intellectual level of modern society without which it will fail to adjust to new realities. It is essential not to teach longer but to teach better and in a way radically different from the old (that is when many of the bright Petrov's ideas will be practical), that is the only chance to survive.

Conclusion

We are facing a difficult voyage through a stormy sea of chaos, from which the structures of the new world can and will emerge. What they will turn out to be largely depends on us. Using the beautiful metaphor of B. B. Rodoman, we should not build the future; instead, we shall cultivate it on the basis of the objective laws of social development. In this voyage, it will be necessary to pass between Scylla of network democracy and Charybdis of healthy authoritarianism, which is strictly necessary for good education. It will hardly be possible to provide a decent education being unable to speak to students honestly and truthfully about their knowledge, abilities, and general level of competence.

However, the game is worth the candle. If successful, this dangerous and hazardous enterprise will give us much more than expeditions of several caravels struggling through storms and winds to uncharted lands, clashing with hostile aborigines, suffering great losses. As a result, the future world will be favourable for thinkers, people who are not obsessed with material consumption, passionate about their practical activity or theoretical issues but critically analysing any information, and therefore not suitable for manipulation. We are not destined to live in such a world, but if we this future to come, we should spare no effort to teach and train "young navigators of the future storm".

References

Akaev AA, Akaeva AA (2011) Vyzovy global'nogo demograficheskogo perekhoda i neotlozhnost' strategicheskikh reshenii (Challenges of global demographic transition and the urgency of strategic decisions). Vek globalizatsii. http://www.socionauki.ru/journal/articles/132577/. Accessed 7 Apr 2017 (in Russ.)

Gorkin AP, Lipets YuG (eds) (2003) Razmeshchenie proizvodstva v rynochnoi srede. Iz trudov B. N. Zimina (Placement of production in a market environment. From the works of VN. Zimin). https://drive.google.com/file/d/0BhZcFCjs36CWEZkeDVkR1NMMVU/view. Accessed 7 Apr 2017 (in Russ.)

Kapitsa SP (1999) Obshchaya teoriya rosta chelovechestva. Skol'ko lyudei zhilo, zhivet i budet zhit' na Zemle (The general theory of the growth of mankind. How many people lived, lived and will live on Earth). Moscow, Nauka (in Russ.)

Lipets Y, Pulyarkin VA (2001) Nelineinye protsessy mirovogo razvitiya (Nonlinear processes of world development). Izv RAN Ser geogr 4:31–37 (in Russ.)

Petrov MK (2004) Istoriya evropeiskoi kul'turnoi traditsii i ee problem (The history of the European cultural tradition and its problems) Moscow, «Rossiiskaya politicheskaya entsiklopediya» (ROSSPEN) (in Russ.)

Shuper VA (2015) Society spatial self-organization as a research field and university course. Regional Research Special Issue:34–41

Sintserov LM (2000) Dlinnye volny global'noi integratsii (Long waves of global integration). Mirovaya ekonomika i mezhdunarodnye otnosheniya 5:56–65 (in Russ.)

Slavin B (2017) Progress: Tsifrovaya demokratiya. https://www.vedomosti.ru/opinion/articles/2017/03/31/683513-tsifrovaya-reabilitatsiya. Accessed 31 Mar 2017

The Notion of Path Dependence in the Regional Economic and Geographical Studies

Anastassia L. Kuznetsova

Abstract

The notion of *path dependence*, which is "dependence on the prior development", appeared in economics as part of the evolutionary approach in the 1970s. Since the 1990s, the term has been used in the evolutionary economic geography. *Path dependence*—dependence on the previous development—makes it structural changes in the economy difficult, and overcoming it requires significant efforts. Economic geographers are studying the regional *path dependence* and "lock-in" of development path, shock phenomena and other sources of the regional *path creation*, the regional technological *path dependence*. Comparative analysis of regions showing *path dependence* is actual as well as analysis of the influence of institutional factors (special zones, territories of advanced development, etc.) on the formation and evaluation of *path dependence*. Some possibilities of applying the concept of *path dependence* are shown in this publication on the materials of the Kaliningrad region.

Keywords

Path dependence · Lock-in · Path creation · Economic restructuring ·
The Kaliningrad region

One of the main questions of economic geography is the reasons for spatial heterogeneity of economic processes. The notions pertaining to the nature of economic processes and the development factors play an increasingly important in this field of studies. Therefore. The development of economic geography as a science is closely related to the development of economic theory and the notions and ideas arising in its context. Path dependence belongs to a group of such ideas.

A. L. Kuznetsova (✉)
Institute of Environmental Management, Urban Development and Spatial Planning,
Immanuel Kant Baltic Federal University, Kaliningrad, Russia
e-mail: anastazija@yandex.ru

© Springer Nature Switzerland AG 2020   11
G. Fedorov et al. (eds.), *Baltic Region—The Region of Cooperation*,
Springer Proceedings in Earth and Environmental Sciences,
https://doi.org/10.1007/978-3-030-14519-4_2

The notion of *path dependence* has emerged in the frame of the evolutionary approach that appeared in the 1970s in opposition to the then prevailing neoclassical economics (Martin 1999).

Path dependence means that history matters and that current state of affairs depends on the historical development of a territory and is a result of the previous development path.

A. Auzan notes that the emergence of the theoretical background for path development analysis goes back to D. North who received the Nobel prize in economics in 1993 (Auzan 2015). According to D. Horth, path dependence means that decisions once made are difficult to cancel.

The most general definition of the path dependence notion was given by P. David, the founder of the concept: "path dependence is a process of changes in which small events, including those purely dominated by chance or luck, of the process's own history, leads to an eventual outcome which might not necessarily be the most efficient one. However, the impact is persistent and enduring and, therefore, leads to the state of lock-in" (David 1985).

The hypothesis made by North and David has created frameworks for the understanding of how and why economic development path differ, what reasons are for choosing the path and how it is possible to shift from one path to another. In particular, the consequence of the hypothesis is a rebuttal of the thesis that development is inevitable and that efficiency and profit maximization determine success. Otherwise, the concept of path dependence allows us to study how institutional innovations become possible and why institutional innovations are not always successful.

However, the notion of path dependence is not just a claim or a proof that 'history matters'. The notion has already been formulated and there is a theory that describes the 'behavior' of these systems. The theory identifies the reasons for path dependence, some of which are enumerated below:

(1) institutional complementarity;
(2) economies of scale;
(3) low value of local assets;
(4) increased returns of technological development;
(5) imperfection of information (Shiryaev 2013).

Since the mid 1990s research works on path dependence have spread from natural sciences and economics to geography, regional development and planning, other social sciences.

Economic geography deals with uneven spatial distribution of economic activity. The evolutionary approach rests on historic processes that form economic structures in space. From the evolutionary paradigm point of view, time and space are intrinsically linked. Thus, economic geography is inseparable from economic growth because spatial patterns emerge from economic processes laid down in the past.

New concepts and instruments of spatial analyses that economists have started to use led to what R. Martin named 'geographical turn' in economics. Following

Krugman, economists discovered economic geography for their research (Krugman 1991). This has entailed increasing diffusion of both sciences and renewed the interest to economic geography. During the last two decades, economic geographers have started to use such evolutionary concepts and metaphors as path dependence, lock-in and co-evolution.

Evolutionary economic geography aims at studying the emergence of spatial particularities of regional specialization, the achieved development level (routines in the economic sense) and its changes. It is especially interesting to use the analyses of the emergence and diffusion of new regional competences (specialization, technology etc.), and the mechanisms of successful competences replication. This change in the approach reflects a more general transition of economic geography from the spatial analysis of capital assets to the analysis of organizational competences, and their anchoring in the local and global economy (Frenken and Boschma 2007).

Most economic geographers employ the concept of path dependence to explain why changes take a particular direction, and how regional industrial patterns vary or fail to change. Research works on regional path dependence focus on the following areas (Henning et al. 2013):

(1) regional path dependence and lock-in;
(2) disruptive events and regional path dependence;
(3) regional technological path dependence and a branching process.

A comparative analysis of regional path dependence, the influence of institutional factors (special economic zones, etc.) on regional path dependence are popular areas of research.

The evolutionary approach in the Russian economic and historical studies gained attention during the last decade. The most impressive research works were written by the scholars from the Russian Academy of Science (V. Polterovich), M. Lomonosov Moscow State University (A. Auzan) and State University High School of Economics (R. Nureev). Economic studies in path dependence theme are mainly concentrated on questions of institutional traps of the Russian economy and socio-cultural aspects of economic transformations in Russia. From the beginning of 2010, a number of articles on path dependence have been written by geographers from Moscow State University (A. Fetisov) under the general title "evolutionary geography" (Fetisov 2011).

Their works are especially important for the Kaliningrad region, Russia's exclave in the Baltic Sea region, as the economy of the region requires serious changes due to continuously changing external and internal factors of its development. The prevailing regional specialization makes this process difficult.

A retrospective analysis shows that the of the Kaliningrad region has been subjected to restructuring several times during a short historical period of its development due to changes in the external environment (Gareev et al. 2005; Gimbitskii et al. 2014; Gareev 2014). Every time the region had to overcome its path dependence with significant difficulties.

I. In the second half of 1940s, at the time of the regional economy formation, the main difficulties were related to the necessity of transformation of market (capitalist) economy, which used to be part of Germany's economy, into centralized administrative (socialistic) economy of the Soviet Union. In the new conditions, many industries remained economically effective only because of the then existing industrial assets (the pulp and paper industry, mechanical engineering and machine building). Their economic effectiveness was beyond any doubt. Small-scale settlement system corresponded to a network of small private agricultural enterprises and small towns. This system existed during the Soviet period. Even now small settlements and towns prevail in the region.

II. During 1950–1970—the period of the region's extensive growth through the creation of new industries—a strong maritime and food processing complex was formed and several precision machinery enterprises were built. The agriculture of the region (dairy and meat production, pig and poultry farming, and the production of fur) was rather effective for the USSR. Land management played an auxiliary role compared to the production of milk, meat and vegetables for the population of the region. Kaliningrad sea ports were rapidly developing. The role of the region as a centre for recreation and tourism was increasing. Another direction of the region's specialization was defense. It was during this period that a big navy base was built and a large number of troops were deployed in the region.

III. Since the second half of the 1970s an attempt to reform Russia's economy in general and that of the region, in particular, was made. The reforms were aimed at raising labour productivity and stimulating economic growth. However, the path dependence effect was still strong because of the high costs of the earlier built capital assets, the structure of the labour market and the professional qualification of the labour force, etc. the main reason was the tradition to develop in an extensive way, to create new jobs that didn't differ a lot from those that already existed. New machines and equipment replaced obsolete assets extremely slowly. The administrative command economy and lack of innovations played an extremely negative role. Attempts to develop the economy of the Kaliningrad region intensively failed.

IV. The second half of the 1980s was a period of the so-called *perestroika*—an attempt to modernize socialism and among other things to introduce a decentralized market economy. During this period, path dependence affected USSR as a whole. Opposition to the old state order was so strong that it led to the collapse of the Soviet Union. The Kaliningrad region became Russia's exclave on the Baltic.

V. The 1990s were characterized by a global qualitative change of the previous path. This change turned out to be even more profound for the Kaliningrad region than for Russia as a whole. There was very little left from the former economic specialization of the region. The most radical changes were caused by the Special Economic Zone regime in the Kaliningrad region and the process of privatization particularly regarding capital assets of the maritime industry (sea vessels were privatized and sold or re-registered under a flag of convenience). The reform of the political system facilitated the formation of a new regional economy.

VI. Since the end of the 1990s, a new economy of the region has been forming based on the effect of the Law "On the Special Economic Zone in the Kaliningrad Region" (1996). Industries based on partial import replacement have been created; they use imported parts or semi-finished goods to produce new products to be exported to other Russian regions. A radical reform of the previous economic system was an indispensable condition for that wiping out the previous path dependence effect.

VII. The adoption of a new law on the Special Economic Zone (2006) was aimed at attracting large investments and a new restructuring of the regional economy. Extending the period of custom preferences allowed Kaliningrad exporters to sell their goods in other parts of Russia enjoying the benefits of the Special Economic Zone. Import-based goods having low added value were an illustration of a new path dependence: the previous form of the economy would not be coming back. That is why when the custom preferences regime was cancelled on April 1, 2016, the Russian Government had to allocate considerable sums to compensate for the losses.

VIII. In 2017, many enterprises, which appeared during the Special Economic Zone of 1996, still existed. Political and economic relations between Russia and Western countries worsened. Russian countermeasures to Western sanctions have negatively affected the industries processing import agriculture produce (especially meat). But path dependence is still visible: it makes it more difficult to form a new economy, which is less dependent on import and logistics costs and is more innovation-based and less energy intensive. By the end of 2017, the State Duma adopted changes to the Law on the Special Economic Zone to stimulate new industries. There is a new strategy of the social and economic development of the region that can lock-out the regional dependence path and bring in new economic patterns.

References

Auzan AA (2015) «Effekt kolei» . Problema zavisimosti ot traektorii predshestvuyushchego razvitiya – evolyutsiya gipotez ("The effect of the rut". The problem of dependence on the trajectory of the previous development is the evolution of hypotheses). Vestn Mosk Univ Seri 6 Ekon 1:3–17 (in Russ.)

David PA (1985) Clio and the Economics of QWERTY. Am Econ Rev 75(2):332

Fetisov AS (2011) Ot istoriko-geograficheskogo ocherka v ekonomiko-geograficheskoi kharakteristike strany k evolyutsionnomu stranovedeniyu (From the historical and geographical outline in the country's economic and geographic characteristics to the evolutionary geography). In: Voprosy ekonomicheskoi i politicheskoi geografii zarubezhnykh stran. Moscow, pp 83–94 (in Russ.)

Frenken K, Boschma RA (2007) A theoretical framework for economic geography: industrial dynamics and urban growth as a branching process. J Econ Geogr 7(5):635–649

Gareev TR (2014) The early evolution of the special economic zone in Kaliningrad oblast. Crisis management challenges in Kaliningrad. Taylor & Francis Ltd, Farnham, pp 19–53

Gareev TR, Zhdanov VP, Fedorov GM (2005) Novaya ekonomika Kaliningradskoi oblasti (New economy of the Kaliningrad Region). Vop Ekonomiki 2:23–29 (in Russ.)

Gimbitskii KK, Kuznetsova AL, Fedorov GM (2014) Razvitie ekonomiki Kaliningradskoi oblasti: etapy restrukturizatsii (Development of the economy of the Kaliningrad region: stages of restructuring). Baltic Region 1(19):56–71 (in Russ.)

Henning M, Stam E, Wenting R (2013) Path dependence research in regional economic development: cacophony or knowledge accumulation? Reg Stud 47(8):1348–1362

Krugman P (1991) Increasing returns and economic geography. J Polit Econ 99(3):483–499

Martin R (1999) Critical survey. The new 'geographical turn' in economics: some critical reflections. Camb J Econ 23:65–91

Shiryaev IM (2013) Zavisimost' ot predshestvuyushchego puti razvitiya i sozdanie puti razvitiya kak vazhneishie kontseptsii v evolyutsionnoi ekonomike (Dependence on the previous path of development and creation of a path of development as the most important concepts in the evolutionary economy). J Econ Regul 4(3):103–112 (in Russ.)

Topical Issues of Cross-Border
Cooperation in the Baltic Sea Region

Transboundary Clustering in Russia's Baltic Coastal Zones Amid Geopolitical Turbulence

Alexander Druzhinin⬡

Abstract

The gravitation of the economy and population towards seacoasts is a universal trend, which is turning marine environments not only in hotbeds of environmental problems but also into arenas for active transnational economic collaborations. In this study, I address economic integration in the Baltic region—particularly, in its Russian segment—and the effect of today's geopolitical turbulence on transboundary clustering. The methodological and theoretical framework for this research spans the theories of globalisation, economic regionalism, and the 'maritime factor' in transboundary integration and clustering. In this work, I present the Baltic region as an aquatic-terrestrial international socioeconomic and environmental-economic system, the development of which was initiated by the rapid European integration in the post-Soviet period. The growing geoeconomic influence of China, India, and other countries of South and East Asia and the geopolitical divergence between Russia and the West shape the new reality of the second decade of the 21st century. I demonstrate that, in these conditions, Russia's Baltic areas (the Leningrad and Kaliningrad regions and Saint Petersburg) retain their potential for transboundary cooperation and transboundary clustering. I emphasise that a central priority is a geoeconomic diversity of emerging and mature clusters' connections. To this end, the 'maritime factor' should be used to full extent, which requires the development of seaport infrastructure.

A. Druzhinin (✉)
North-Caucasian Research Institute of Economic and Social Problems,
Southern Federal University, Rostov-on-Don, Russia
e-mail: alexdru9@mail.ru

A. Druzhinin
Immanuel Kant Baltic Federal University, Kaliningrad, Russia

© Springer Nature Switzerland AG 2020
G. Fedorov et al. (eds.), *Baltic Region—The Region of Cooperation*,
Springer Proceedings in Earth and Environmental Sciences,
https://doi.org/10.1007/978-3-030-14519-4_3

Keywords
Transboundary connections · Clusters · Coastal zones · Baltic · Russia · Eurasia

Introduction

Since the late 1980s, when the USSR and its military, political, and economic bloc were on the brink of dissolution, the Baltic Sea coast—including its southeastern and eastern stretches—have been involved in large-scale European integration processes. This gave rise to 'Baltic regionalisation'—the emergence of the 'Baltic region' as an international macroregional whole (Mezhevich et al. 2016). The Baltic regionalisation process intensified in the mid-2000s (Hosli et al. 2009). At the time, Poland, Lithuania, Latvia, and Estonia acceded to the EU, whereas a favourable economic situation was turning Russia's North-West—particularly, Saint Petersburg and the Leningrad and Kaliningrad region—into not only a 'communications corridor' but also a 'development corridor' incorporated into the global economy.

Although creating mostly beneficial conditions for transboundary ties in the coastal areas of Russia's North-West (Druzhinin 2016), the Baltic region has been faced in recent years with multi-aspect external and internal challenges. Committed to common interests and even identities of constituent countries and regions (Fedorov et al. 2012) and characterised by asymmetric economic dependencies, the Baltic region format has partly collapsed and partly transformed. All this has affected the processes of economic clustering.

The Global Economic and Demographic Changes of the Early 21st Century: The Baltic Perspective

Global and Eurasian trends have most significantly affected the realm of geo-economy, which—as experts stress (Dizen 2017)—is no longer the remit of the West. Over the past decade and a half, the Eurasian continent has witnessed an eastward shift in the economic potential with China rising to become a geoeconomic pole, equal to the EU.

As the World Bank's twenty-five-year statistics show, the contribution of both the old[1] and the new[2] 'European West' to the total Eurasian GDP reduced from 51 to 39%, whereas that of China grew from 2.5 to 23.5% (Table 1).

---

[1]The European states that were members of NATO and the EU before 1991, as well as Austria, Andorra, Cyprus, Malta, Monaco, Lichtenstein, Finland, Switzerland, and Sweden.
[2]Albania, Bulgaria, the Czech Republic, Croatia, Estonia, Hungary, Latvia, Lithuania, Poland, Romania, Slovakia, Slovenia.

Table 1 The contribution of macroregions, associations, and states to the total Eurasian GDP (at the official conversion rate), %

	1992	2000	2013	2015
Eurasian Economic Union	2.82	1.41	5.18	3.37
of which Russia	2.57	1.25	4.52	2.83
'old European West'	49.61	44.39	38.33	36.36
'new European West'	1.66	2.06	2.89	2.74
China	2.58	5.84	19.45	23.41
Japan	21.54	22.83	9.94	8.77
India	1.64	2.30	3.77	4.46

Source Prepared by the author based on World Bank data

Amid general geopolitical changes, the global positioning of the Baltic metaregion is changing too. According to approaches found in the literature (Fedorov et al. 2012), the Baltic region includes Denmark, Latvia, Lithuania, Finland, Sweden, Estonia, and certain areas of Germany, Poland, and Russia. In 1992, these countries accounted for 12.8% of the gross world product (at the official conversion rate). In 2000, their contribution was estimated at 8.9%, in 2013, at 10.4%, and, in 2015, at 8.5%. Note that, if Russia is taken out of the calculation, the proportion of the other Baltic countries will steadily decrease from 11.0% in 1992 through 8.0% in 2000 and 7.4% in 2013, to 6.7% in 2015. The general trend is explained by the slow economic growth rates observed in Denmark, Sweden, and Finland and the economic performance of Germany—the regional behemoth that has been steadily its position as a major contributor to the gross word product since the 1990s.

Although remaining in a privileged position within the global core/periphery system, not only the Baltic region recreates the existing cross-country and cross-region socioeconomic differences but it is also turning into one of many major geoeconomic hubs of the emerging multi-polar world economic system.

Structural changes are accompanied by demographic processes that are transforming the Baltic States into an area of not only relative (in comparison to the global dynamics) but also absolute depopulation. According to the UN data, only in 2010–2015, the annual population decline rate reached 0.2% in Germany, 0.1% in Estonia, 0.4% in Lithuania, and 0.4% in Latvia. Poland's population ceased to grow. Only the Baltic region's Nordic segment—Denmark, Finland, and Sweden—is witnessing a population increase (0.3–0.6%) that is backed by positive net migration. However, in the most socioeconomically developed coastal regions of Germany (with the exception of Mecklenburg-West Pomerania) and Poland, the population is growing despite the overall negative demographic trends (Fedorov et al. 2017). A similar situation is observed in the Russian sector of the Baltic (Druzhinin 2017a, b), which is explained by the attractiveness of the Saint Petersburg and Kaliningrad agglomerations. Nevertheless, since as early as the mid-20th century, the proportion of Europe (which includes Russia, according to the UN) in the total Eurasian population has been steadily decreasing. Its contribution fell from 27 to 14% in 1950–2015. At the same time, the demographic potential of the countries of South, East, and West Asia has been growing.

The transition to a multi-polar system is accompanied (see Strategy of economic security… 2017) by growing geopolitical instability, the unsustainable development of the world economy, and intense global competition.

Rising Tensions Between Russia and the West: The Emergence of the 'Baltic Frontier"

The ambition to not only 'contain' Russia (Brzezinski 1998), as the tradition is, but also to stop China (Brune and Guichard 2012) and to prevent further undesirable changes in the global power equation is a main driver of the recent eastward and south-eastward expansion of the Euro-Atlantic structures. Throughout the past years, by 'enlarging' the EU and NATO and by organising 'synergies' and 'partnerships' of various types, the West has been 'marking territory', reformatting and readjusting national economies to its own benefit, and trying to transform local identities. Since 2007–2008, the ongoing 'Westernisation' of the post-Soviet space has been provoking expressions of concern from the Russian Federation. This cannot but contribute to the country's anti-Western sentiment. All this is turning the Baltic region and, primarily, its southern and eastern periphery into a significant geopolitical frontier and an area of tension, which is drawing the attention of global and regional actors. Such a situation translates into growing economic risks for Russia and the neighbouring Baltic countries and into the primacy of geopolitics over economics.

Since spring-summer 2014, the 'frontier' function of the aquatic and terrestrial parts of Russia's Baltic segments has increased amid the Ukraine and Syria crises. The West's negative reaction to Russia's involvement in the events led to sanctions and counteractions. Against the background of divergence between Russia and the other Baltic countries and 'frozen' regional cooperation, the Baltic region reassumed the bi-structural geopolitical architecture, which it had had until the late 1980s, and re-established itself as a key fragment of the military, political, and ethnocultural barrier spanning a broad, partly 'blurred' arc from the Arctic to the Middle East. At the same time, the existing transboundary ties started to decay and transform. The new conditions for the functioning of coastal border regions—in particular, Russia's Baltic exclave of Kaliningrad—became apparent.

The Potential for Transboundary Cooperation in the New Geopolitical Reality

In the new geopolitical reality, there is still potential for transboundary cooperation in Russia's Baltic regions. There are several reasons for that.

Firstly, despite the trend-dependent diversification of Russia's economic ties, the EU member states remain the country's major partner. Today, the EU accounts for 42% of Russia's international trade. The significance of the EU as an economic partner was symptomatically emphasised in the Framework for the Foreign Policy of the Russian Federation, which was approved in November 2016. Russia's geoeconomic specialisation, which developed in the post-Soviet period, remains intact. Moreover, amid the growing cross-country (cross-bloc and cross-civilisation) competition and confrontation and turbulence in the global raw materials markets, Russian exports are not falling but, on the contrary, growing (Druzhinin 2018). The tonnage handled by the Russian Baltic seaports is increasing. The Leningrad and Kaliningrad regions Saint Petersburg are still serving as major channels for imports in the country.

Paradoxically, the 'barrier' function of Russia's northwestern coastal borderlands co-exists with the 'communicative' function in terms of geo-economy, transport, and logistics. Moreover, the latter function is becoming increasingly important, being supported by the Russian-driven Nord Stream 2 project. Caused by global processes, the geopolitical divergence does not cancel out the significance of geoeconomic neighbourhood either for Russia or for its partners in the Baltic region. Nor does it downplay the ethnocultural, sociodemographic, environmental, economic, and geopolitical challenges. Even amid cooling relations between Russia and the other countries of the Baltic region, there are beneficial conditions for transboundary clustering in the key coastal zones of Russia's North-West.

Another important factor and incentive for stronger transboundary ties relates to the crisis and stagnation of the Russian economy. In terms of the GDP generated by the economy, Russia has been thrown back to the 2007 levels (at the official conversion rate). This is contributing to a growing economic and social gradient between Russia's coastal and border regions, on the one hand, and their counterparts in the neighbouring Baltic countries. For instance, as of the beginning of 2014, the average gross nominal salary in the Kaliningrad region was equivalent to USD 800. A year later, it did not exceed USD 450, which translated in a significant gap between the incomes of the residents of Kaliningrad and the neighbouring Polish voivodeships. As of 2017, the average gross nominal salary in the Warmian-Masurian voivodeship was USD 940 (Statistical Office 2016). These circumstances contribute to the traditional spatial transboundary socioeconomic gradients. Their growing tangibility is supported by the elements of the so-called fourth industrial revolution. Characterised by a fusion of robotics, virtual economy, the Internet of Things, 3D-printing, and artificial intelligence, it devaluates the existing infrastructure and human capital, destroys economic clusters, and creates incentives for new cluster initiatives that turn vast areas into peripheries. Therefore, despite the withdrawal of the UK from the EU, there are reasons to believe that the European integration is far from complete and that the EU is still perceived as (and probably is) the most mature and successful integration association.

Thirdly, it is necessary to take into account the Chinese factor—which is growing in importance even in the Baltic area—and the infrastructure, transport, production, and logistics projects with Chinese participation. In this context,

Russia's northern and southwestern transport and communications corridors, which have been modernised over the past two decades, will remain major thoroughfares under almost any geostrategic scenario. Moreover, these corridors will stimulate and support the development of transboundary clustering in Russia's coastal areas. The prevalence of the EU—China geoeconomic dyad, the role of seas and coastal infrastructure in carrying the traffic between them, and probable externalities associated with the localisation of Chinese interests in the Baltic region constitute another strategically significant factor of rapid clustering in the Saint Petersburg coastal area and the exclave Kaliningrad region.

Transboundary Clustering Priorities in Russia's Coastal Zones in the Baltic

The reliance on the growing potential of the Big Eurasian Partnership is the basic but not the only aspect of the transboundary clustering in Russia's Baltic coastal zones. This being said, the Big Eurasia project should be perceived as the only feasible option for Russia (Dizen 2017). Amid the growing geopolitical and geoeconomic turbulence and associated economic risks, it is strategically important to treat the *stability* of clusters and their resistance to both expected and spontaneous changes in cross-country relations as a priority when developing approaches and measures to support transboundary clustering. (Expected changes are caused by fluctuations in the world and the partners and interests altering with the economic situation, whereas spontaneous ones are politically motivated.)

Treating stability as a basic priority requires, in particular, that clusters develop diversified economic ties. In such a case, significant deterioration of relations with a neighbouring state or a group of countries will not lead to a catastrophe. For coastal zones, it is vital that a cluster's diversified ties engage port facilities and maritime logistics, i.e. take advantage of the maritime factor and the port infrastructure. It is strategically important not to overlook the coastalisation component of transboundary clusters. In most cases, clusters emerge either within maritime industries (shipbuilding, coastal and maritime tourism) or 'in collaboration' with port facilities.

The stability of transboundary clusters should rest on the potential orientation of the 'Russian segments' to both the domestic market and the markets of neighbouring states, as well as of those accessible by sea. Another source of stability is the marketing flexibility of cluster participants. However, a situation when a transboundary cluster's core or 'profit centre' is beyond Russian jurisdiction is considered the last expedient. Such a cluster will require a transformation into a double- or multiple-core *Russia-centred* structure that will generate a sufficient socioeconomic and innovative effect for Russian coastal territories.

Conclusion

The 'natural way', the components of transboundary cluster localise within the major 'development corridors' of the leading and most dynamic coastal agglomerations that are attractive for businesses, investment, and migrants. This is well in line with the logic of the post-Soviet, market, and capitalist spatial organisation of society. However, there is a pressing need to expand and diversify the clustering area by involving semi-periphery and periphery territories, using regional policy tools. In today's geopolitical situation, an equally important factor is public support for cluster initiatives in coastal and border regions with special economic conditions. In Russia's Baltic segment, such a case is the Kaliningrad region—the second largest arena for cooperation between Russia and the Baltic region, after Saint Petersburg. Being absolutely necessary, such cooperation has significant potential and attractive prospects.

Acknowledgements The study was supported by a grant from the Russian Science Foundation, 18-17-00112, 'Ensuring the Economic Safety of Russia's Western Borderlands amid Geopolitical Turbulence'.

References

Brune A, Guichard J-P (2012) Geopolitika merkantilizma: novyi vzglyad na mirovuyu ekonomiku i mezhdunarodnye otnosheniya (Geopolitics of mercantilism: a new look at the world economy and international relations). Novyi khronograf, Moscow (in Russ.)

Bzhezinskii Z (1998) Velikaya shakhmatnaya doska. (Gospodstvo Ameriki i ego geostrategicheskie imperativy) (Great chess board. (Dominance of America and its geostrategic imperatives). Mezhdunarodnye otnosheniya, Moscow (in Russ.)

Dizen G (2017) Rossiya, Kitai i «balans zavisimosti» v «Bol'shoi Evrazii» (Russia, China and the "balance of dependence" in "Greater Eurasia"). Valdaiskie Zap 63:3–12 (in Russ.)

Druzhinin AG (2016) «Morskaya sostavlyayushchaya» rossiiskoi obshchestvennoi geografii: traditsii i novatsii ("Marine component" of Russian social geography: traditions and innovations). Izv RAN Ser Geogr 6:7–16 (in Russ.)

Druzhinin AG (2017a) Talassoattraktivnost' naseleniya v sovremennoi Rossii: obshchestvenno-geograficheskaya eksplikatsiya (Thalassoattractiveness of the population in modern Russia: socio-geographical explication). Baltic Reg 2:28–43 (in Russ.)

Druzhinin AG (2018) «Odin poyas — odin put'» : vozmozhnosti dlya regionov zapadnogo porubezh'ya Rossii ("One belt—one way": opportunities for the regions of the western neighborhood of Russia). Baltic Reg 2:39–55 (in Russ.)

Druzhinin AG (ed) (2017b) Transgranichnoe klasteroobrazovanie v primorskikh zonakh Evropeiskoi chasti Rossii: faktory, modeli, ekonomicheskie i ekisticheskie effekty (Transboundary cluster formation in the coastal zones of the European part of Russia: factors, models, economic and athlete effects). Southern Federal University Publication, Rostov-on-Don (in Russ.)

Fedorov GM, Mikhailov AS, Kuznetsova TYu (2017) Vliyanie morya na razvitie ekonomiki i rasseleniya stran Baltiiskogo regiona (The influence of the sea on the development of the economy and the resettlement of the countries of the Baltic region). Baltic Reg 2:7–27 (in Russ.)

Fedorov GM, Zverev YuM, Korneevets VS (2012) Rossiya na Baltike: 1990–2012 gody (Russia in the Baltic: 1990–2012). Immanuel Kant Baltic Federal University Publication, Kaliningrad (in Russ.)

Hosli M, Mattila M, Uriot M (2009) Voting behavior in the council of the European union after the 2004 enlargement. In: EUSA, Conference Paper, Los Angeles, 23–25 April 2009

Mezhevich NM, Kretinin GV, Fedorov GM (2016) K voprosu ob ekonomiko- geograficheskoi strukturizatsii Baltiiskogo regiona (On the question of the economic and geographical structuring of the Baltic region). Baltic Reg 8(3):15–29 (in Russ.)

Statistical Office in Olsztyn (2016) Raport o sytuacji społeczno-gospodarczej województwa warmińsko-mazurskiego w 2016 r. http://olsztyn.stat.gov.pl/publikacje-i-foldery/warunki-zycia/raport-o-sytuacji-spoleczno-gospodarczej-wojewodztwa-warminsko-mazurskiego-w-2016-r-,2,6.html. Accessed 3 July 2017

Strategy of economic security of the Russian Federation for the period until 2030 (2017) http://www.garant.ru/products/ipo/prime/doc/71572608/. Accessed 29 June 2017 (in Russ.)

On the Directions and Prospects of Cross-Border Cooperation Between Russia and the EU Countries in the Baltic Region

Gennady M. Fedorov

Abstract

The Baltic Sea is being gradually enclosed by an international macroregion—an arena for rapidly developing transnational and transboundary cooperation. The Baltic macroregion includes eight members of the European Union and one non-EU country—Russia. The EU pursues the policy aimed at development of socioeconomic ties between the Baltic EU member states and their regions. Russia and its Baltic territories are taking an active part in international cooperation. In this article the forms of such cooperation are considered. Current political tension between Russia and the West is complicating mutually beneficial collaboration. The data presented in the article demonstrates the declining volumes of trade between Russia and other countries of the macroregion in 2015–2016 and a growth therein in 2017. In 2010–2016, the level of investments from the EU Baltic countries in Russia decreased. In 2016, Russian investment in these states declined too. Social contacts reduced in intensity, although collaboration in major areas continued. The year 2017 saw some positive changes in bilateral trade. Projects aimed at boosting cross-border cooperation between Poland and Russia and Lithuania and Russia are being devised in 2018 which gives grounds for hope that the EU—Russia collaborations in the Baltic will pick up steam.

Keywords

Baltic region · Transboundary cooperation · International trade volume · Foreign investment · Russia · European union

G. M. Fedorov (✉)
Institute of Nature Management, Spatial Development, and Urban Planning,
Immanuel Kant Baltic Federal University, Kaliningrad, Russia
e-mail: gfedorov@kantiana.ru

Introduction

Spreading rapidly since the mid-20th century, globalisation entailed international regionalisation. The Baltic region is one of the world leaders in this regard—the Baltic Sea is being gradually enclosed by a transnational macroregion. Nine countries on it shores are creating transboundary meso- and microregions. This process is supported by international cooperation across almost all the areas of production and social contacts. Regionalisation is associated with greater international competitiveness of the countries and regions involved.

In this article, I summarise findings obtained by geographers, economists, and political scientists at the Immanuel Kant Baltic Federal University. Many of the studies considered are collaborations between researchers from Russia and the other countries of the region. Most of the findings were published in the University's Scopus and WoS-indexed *Baltic Region* journal. Some of the results were discussed at international conferences. In this work, I examine forms of cooperation that contribute to a stronger Baltic region and emphasise resultant benefits to all the parties involved.

Forms of International Cooperation in the Baltic

In the 1990s, after the dissolution of the USSR, and even in the early 2000s, many Russian and international experts expected the Baltic region to become an arena for active cooperation among all the local countries, including Russia. Indeed, all kinds of ties were developing in the region. The Kaliningrad region devised a strategy that envisaged it as a territory of international collaborations (Government of the Kaliningrad Region 2003).

The forms of Baltic cooperation that involve—to a degree—Russia's regions and their municipalities, economic entities, and NGOs are as follows:

– membership in the Council of Baltic Sea States. Established in 1992, the Council brings together 9 countries bordering the Baltic Sea, as well as Norway, Iceland, the European Commission, and eight observer states (The Council of the Baltic Sea States 2017);
– collaborations within the Visions and Strategies around the Baltic Sea (VASAB) programme (Visions and Strategies around Baltic Sea 2017);
– Russia–EU projects within transboundary cooperation programmes. The 2007–2013 Russia–EU projects—'Estonia—Latvia—Russia', 'Lithuania—Poland—Russia', 'Karelia','Colarctic', 'South-East Finland—Russia'—included 200 projects aimed to boost small and medium businesses, to support local cultures and traditions, and to prepare recommendations for the development of border regions. Over 50 large infrastructure projects focused on the equipment of border checkpoints, the development of transport networks and tourism infrastructure, and finding solutions to environmental problems.

In 2018, local actors are preparing project applications within the Russia-EU cross-border cooperation programme for 2014–2020. Although the previous programme included both bi- and tri-lateral project, the new one encompasses only bilateral ones, namely, Karelia, Colarctic, Russia—Latvia, Russia—Lithuania, Russia—Poland, Russia—Estonia, and Russia—South-East Finland (Russia-EU cross-border cooperation programs) (Russia-EU Cross-Border Cooperation Programs 2017). Moreover, Russia is a participant in the Interreg Baltic Sea Region transboundary cooperation programme;

– euroregions with Russian participation (Fedorov 2012).

Popular forms of transboundary cooperation in Europe, euroregions are associations of municipalities and other administrative units that coordinate their actions based on cooperation agreements in different areas within regional competences. In the 1990s-early 2000s, seven euroregions with Russian participation were created (Table 1).

– local border traffic. Russia has concluded agreements with Norway and Latvia. The agreement with Poland, which was in effect in 2012–2016, was suspended on the initiative of the national Polish authorities, despite protests from the general public and the authorities of border districts. The agreement boosted the transboundary traffic. In 2014, 4.7 million crossings were carried out within the local border traffic regime (Gumenyuk et al. 2016);
– bilateral and multilateral collaborations among educational institutions and research, cultural, and sports centres. Unique to the Baltic region, the 'Baltic University' education programme was initiated by Uppsala University, Sweden (Baltic university programme 2017). The programme coordinates environmental, and international cooperation, and other actions of universities situated within the catchment area of the Baltic Sea rivers;
– international cooperation and workshops on a range of problems of mutual interest.

Table 1 Euroregions with Russian participation in the Baltic macroregion

No	Euroregion	Founded	Participating countries
1	Neman	1997	Poland, Lithuania, Belarus, Russia
2	Baltic	1998	Poland, Lithuania, Sweden, Denmark, Russia
3	Saule	1999	Lithuania, Latvia, Russia, Sweden
4	Karelia	2000	Finland, Russia
5	Łyna-Ława	2003	Poland, Russia
6	Šešupė	2003	Russia, Lithuania, Poland, Sweden
7	Pskov-Livonia	2003	Russia, Estonia, Latvia

Source Zaukha et al. (2008)

International Trade

The pillars of a consolidated Baltic region are international economic ties and, particularly, trade and foreign investment.

The Baltic region states are Russia's major trade partners. In 2017, they accounted for 16.8% of the country's international trade. Germany's contribution is estimated at half of this proportion (8.6%). For a long time, the country was Russia's largest trade partner. However, today, it has lost its position to China (14.9%). In terms of bilateral trade with Russia per capita, Latvia ranks first, Finland second, and Estonia third among the Baltic region states. Germany outperforms only Poland and Sweden (Table 2).

Unfortunately, the confrontation between Russia and the West, which has been taking place over the past decade, did not contribute to economic cooperation. Both parties have sustained significant loses. Figure 1 shows a steep reduction in Russia's bilateral trade with all the other Baltic region countries in 2014–2016.

The year 2017 saw a growth in Russia's trade with the Baltic region countries (Fig. 2). The most significant increase was observed in bilateral trade with Denmark. However, none of the countries could regain the level of 2014. Moreover, Russia's bilateral trade with Latvia and Estonia shrank further that year.

A reduction in the bilateral trade between Russia and the other Baltic region states in 2014–2016 is largely explained by the plummeting rouble—a result of falling oil prices. Other reasons for such a reduction are the worsening relations between Russia and the EU (and the West in general), western sanctions, and Russian countersanctions. Thus, in 2013–2017, the contribution of the Baltic region countries to Russia's international trade dwindled from 18.6 to 16.8% and that of all the EU member states from 49.0 to 42.2%. Note that, in 1994, the Baltic region countries comprised 23% and, in 2007, 20.6% of Russia's international trade (Rosstat 2017a).

Table 2 Russia's bilateral trade with other Baltic Sea countries

Country	Proportion in Russia's international trade, %	Bilateral trade with Russia per capita, USD
Germany	8.6	0.6
Poland	2.8	0.4
Sweden	0.7	0.7
Denmark	0.7	0.4
Finland	2.1	2.3
Lithuania	0.6	1.2
Latvia	0.9	2.7
Estonia	0.4	2

Prepared by the author based on The Federal Customs Service (2017)

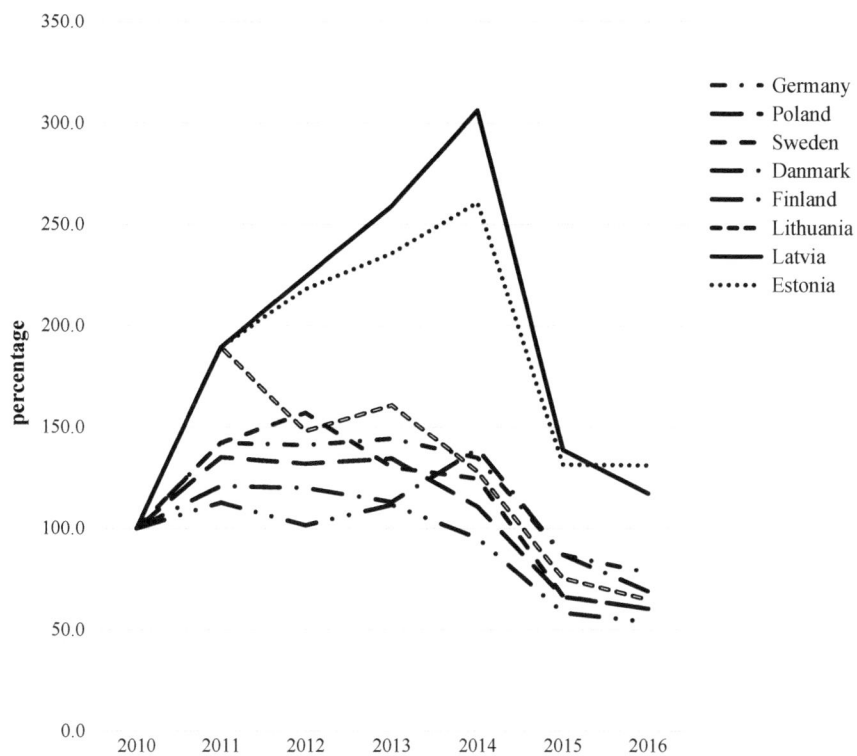

Fig. 1 Dynamics of Russia's bilateral trade with other Baltic region countries, 2010–2016 (% of the 2010 level). Prepared by the author based on The Federal Customs Service (2017)

Foreign Investment

Just like in the case of international trade, the contribution of the EU member states, including those of the Baltic region, in foreign investment in Russia is decreasing. Russian investment in these countries is dwindling too. At the same time, Russia is witnessing increasing investment from China and Japan, with the EU states keeping their leadership position (Evstigneeva 2017).

In 2016, the Baltic region states accounted for half of the increase in the EU's foreign direct investment in the Russian economy. The increase in Russia's direct investment in the region comprised 64% of the increase in their investment in Russia. In terms of the increase in investment in Russia, Sweden ranks first, Finland second, and Germany third among the Baltic region countries. Germany, Denmark, and Sweden saw the greatest rate of increase in investment from Russia (Rosstat 2017b).

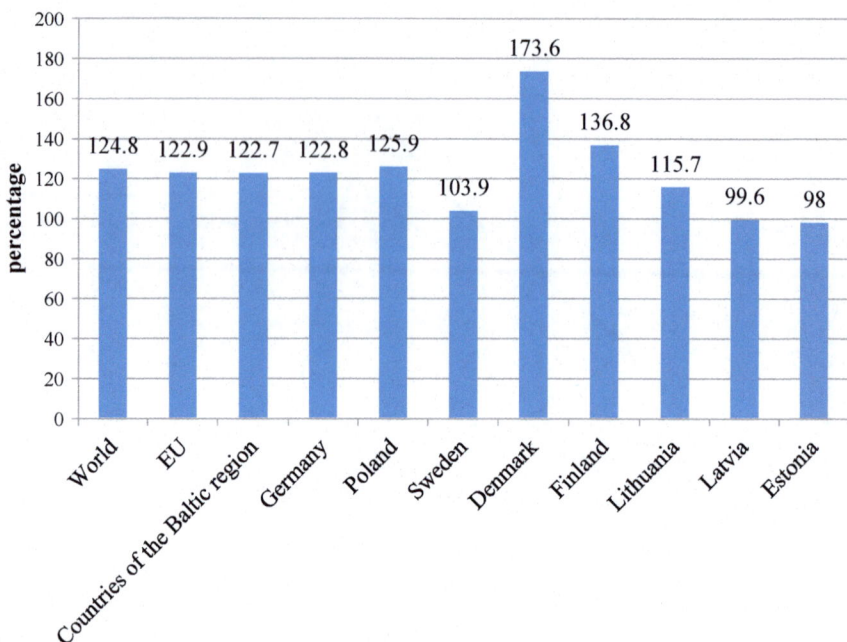

Fig. 2 Changes in Russia's bilateral trade with the Baltic region countries in 2017, % of the 2016 level. Prepared by the author based on The Federal Customs Service (2017)

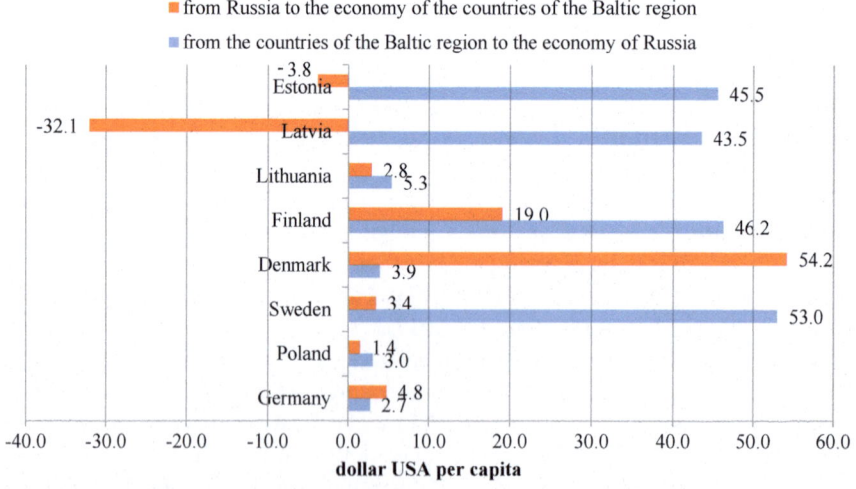

Fig. 3 Mutual investment from Russia in other Baltic region states per capita, 2017. Prepared by the author based on Rosstat (2017b)

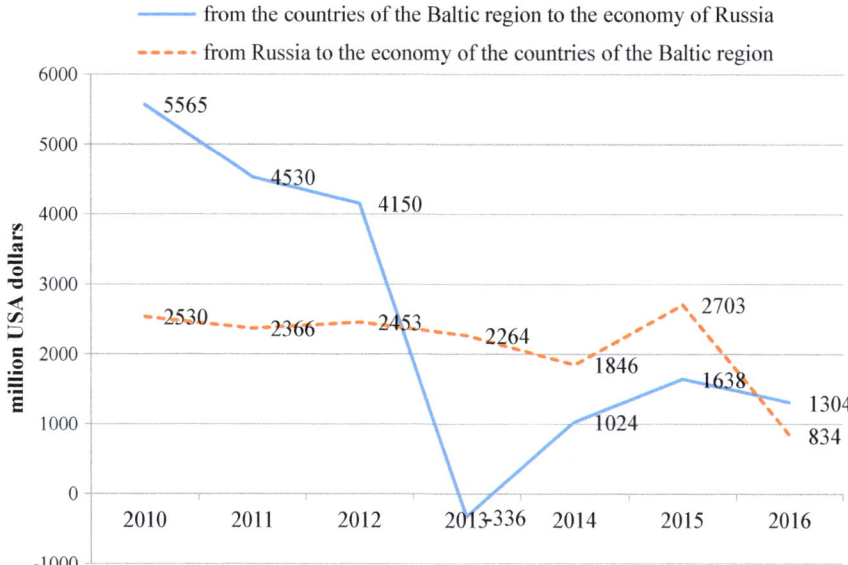

Fig. 4 Dynamics of foreign direct investment in Russia from the Baltic region countries, 2010–2016. Prepared by the author based on Rosstat (2017b)

As to investment in the Russian economy per capita, the 2016 leaders were Sweden, Finland, Estonia, and Latvia, whereas Denmark accounted for the most considerable investment from Russia per capita (Fig. 3).

The worsening political relations between Russia and the EU had a negative effect on mutual investment. However, the investment in Russia from the Baltic region countries started to decrease before 2014, whereas Russia's investment in the region shrank only in 2016 (Fig. 4).

Conclusion

I hope that economic interests and the ambition to become more competitive in the world market will encourage the EU to lift anti-Russian sanctions (which will lead to Russia cancelling its counter-measures) and help the parties to return to the development of mutually beneficial ties. The Baltic region provides a favourable background for intra- and cross-industry clustering, the development of territorial production systems and transboundary regions with Russian and EU participation, and stronger innovation collaborations. A better spatial organisation of the region—the target of the VASAB and Interreg programmes—can contribute to regional competitiveness in the world markets. There is still an untapped potential for mutual investment and joint ventures. Closer cooperation has great prospects in innovations, tourism, education, culture, and sports.

The Baltic is still witnessing the emergence of international regions of different hierarchical levels, including the Baltic transnational regions and transboundary meso- and microregions. Alluring prospects lie in the creation of Russian-Polish-Lithuania region in the South-East Baltic and mesoregions at the Russian borders with Belarus, Latvia, Estonia, and Finland. Within these regions, collaborations among Russian, Polish, and Lithuanian municipalities may give rise to microregions, for instance, a Vistula Lagoon or a Vištytis microregion in the South-East Baltic.

The potential participation of Russia's coastal region in international macro-, meso-, and microregions is taken into account in proposals for the creation of transboundary production systems within the South Baltic and East Baltic 'growth triangles' (Kivikari 2001). In developing the idea of a Tricity (Gdańsk—Gdynia—Sopot)—Kaliningrad bipolar territorial system, which had been proposed by the Polish economic geographer from the University of Gdańsk, Tadeusz Palmowski (Palmowski 2004), Kaliningrad scholars introduced a tripolar model that included Lithuania's Klaipėda (Fedorov et al. 2008). The geographical closeness of Saint Petersburg, Helsinki, and Tallinn gives ground to hope for a similar tripolar model to be developed in the north. The benefits of industrial cooperation may encourage businesses from the three countries to expand and develop mutual connections that will once transform into a tripolar system.

Acknowledgements The study was supported by a grant from the Russian Geographical Society (47. The Russian Baltic: Current Condition, Problems, and Prospects, supervised by Prof. V. M. Razumovsky, Saint Petersburg State Economic University, Saint Petersburg).

References

Baltic University Programme (2017) A regional university network. http://www.balticuniv.uu.se/. Accessed 14 Oct 2017

Council of the Baltic Sea States (2017). http://www.cbss.st/. Accessed 14 Oct 2017

Evstigneeva A (ed) (2017) Pryamye investitsii v Rossiyu iz Azii vyrosli na chetvert' (Direct investment in Russia from Asia rose by a quarter). https://iz.ru/682359/alina-evstigneeva/priamye-investitcii-v-rossiiu-iz-azii-vyrosli-na-chetvert. Accessed 25 June 2018 (in Russ.)

Fedorov GM (2012) Sozdat' «treugol'niki rosta» (Create "growth triangles"). Akkreditatsiya v Obraz 59:36–39 (in Russ.)

Fedorov GM, Zverev Y, Korneevets VS (2008) Rossiya na Baltike: 1990–2007 gody 9 Russia in the Baltic: 1990–2007). RGU im. I. Kanta Publication, Kaliningrad (in Russ.)

Government of the Kaliningrad Region (2003) Strategiya sotsial'no-ekonomicheskogo razvitiya Kaliningradskoi oblasti kak regiona sotrudnichestva na period do 2010 g. (Strategy of social and economic development of the Kaliningrad region as a region of cooperation for the period up to 2010). Kaliningrad (in Russ.)

Gumenyuk IS, Kuznetsova TY, Osmolovskaya LG (2016) Mestnoe prigranichnoe peredvizhenie kak effektivnyi instrument razvitiya prigranichnogo sotrudnichestva (Local border movement as an effective tool for the development of cross-border cooperation). Baltic Reg (1):97–117 (in Russ.)

Kivikari UA (2001) Growth triangle as an application of the northern Dimen-sion policy in the Baltic sea region. Russian-Europe centre for eco-nomic policy. Policy Paper. http://www.etela-suomi.fi/english/pdf/kivikari. Accessed 10 July 2017

Palmovskii T (2004) Novaya baltiiskaya bipolyarnaya model' mezhregional'nogo sotrudnichestva (New Baltic bipolar model of interregional cooperation). Vestn Kaliningradskogo Gosu-darstvennogo Univ Ser Regionovedenie 4:66–75 (in Russ.)

Russia-EU Cross-Border Cooperation Programs. http://economy.gov.ru/minec/activity/sections/mps/programs/. Accessed 26 Oct 2017 (in Russ.)

Rosstat (2017a) Rossiiskii statisticheskii ezhegodnik (Russian statistical yearbook). http://www.gks.ru/wps/wcm/connect/rosstat_main/rosstat/ru/statistics/publications/catalog/doc_1135087342078. Accessed 14 Oct 2017 (in Russ.)

Rosstat (2017b) Rossiya i strany - chleny Evropeiskogo soyuza (Russia and the countries members of the European Union). Moscow (in Russ.)

The Federal Customs Service (2017). http://customs.ru/index2.php?option=com_content&view=article&id=26531:-2017-&catid=51:2011-01-24-16-27-51&Itemid=1977. Accessed 25 June 2018 (in Russ.)

Visions and Strategies around Baltic Sea (2017). http://www.vasab.org/. Accessed 14 Oct 2017

Zaukha Y, Limonov L, Oding N, Fedorov G (eds) (2008) Severo-Zapad Rossii v regione Baltiiskogo morya: problemy i perspektivy ekonomicheskogo vzaimodeistviya i sotrudnich-estva (North-west of Russia in the Baltic Sea Region: problems and prospects for economic interaction and cooperation). RGU im. I. Kanta Publication, Kaliningrad (in Russ.)

The Development of the "Pskov—Pechory—Tartu" Transboundary Tourism and Recreation Region

Andrey G. Manakov

Abstract

The social and economic periphery of the Pskov region (Russia) and the southeastern part of Estonia is in need of new growth poles that would reverse the situation in the border areas. This research seeks to identify the characteristic features of Russia's and Estonia's border areas comprising a transboundary tourism and recreation region. The central method used in this research is the evaluation of primary and secondary factors behind the formation of the transboundary tourism and recreation region. These factors are instrumental in estimating the prospects of transboundary tourism and recreation microregions of different levels. The study has shown that the Pechory district of the Pskov region and Estonia's neighbouring Võrumaa and Põlvamaa districts constitute a transboundary tourism and recreation microregion, which I call "Setomaa". Boasting unique tourism and recreation resources, this transboundary microregion has excellent development prospects. I suggest that a higher-level transboundary microregion that brings together Setomaa and the "core" cities of Pskov and Tartu should be referred to as "Pskov—Pechory—Tartu".

Keywords

Transboundary region · Tourism · Recreation · Estonia · Pskov region

A. G. Manakov (✉)
Department of Geography, Pskov State University, Pskov, Russia
e-mail: region-psk@yandex.ru

© Springer Nature Switzerland AG 2020   37
G. Fedorov et al. (eds.), *Baltic Region—The Region of Cooperation*,
Springer Proceedings in Earth and Environmental Sciences,
https://doi.org/10.1007/978-3-030-14519-4_5

Introduction

In 2004, the Pskov-Livonia euroregion linked the borderlands of Estonia, Latvia, and Russia (the Pskov region) as a continuation of collaborations within the Cross-border Cooperation Council established in 1996. Initially, the euroregion included five Russian, four Latvian, and three Estonian municipalities (Barinov and Kiryushin 2013). This euroregion (which shows features of a "complex" trans-boundary region) gave rise to industry-specific collaborations, in particular, those in tourism and recreation.

Bound by a common tourism resource, a transboundary tourism and recreation region brings together borderland districts of neighbouring countries. It boosts the development of cross-border ties, which translate into collaborations among economic entities. In the Baltic macroregion, there are eight transboundary tourism and recreation mesoregions, one of them connecting Latvian, Estonian, and Russian municipalities (Kropinova 2016).

In my opinion, this is still an emerging mesoregion. However, it includes a highly developed microregion that performs the function of a transboundary entity much more effectively than its counterparts. Straddling the Russo-Estonian border, this transboundary tourism and recreation region includes two major regional centres and the corridor between them. The latter qualifies as a transboundary microregion of the lowest level. Using the names of these centres, one can refer to the macroregion as the "Pskov—Pechory—Tartu" region.

The aim of this research is to describe the development of the "Pskov—Pechory—Tartu" transboundary tourism and recreation region, which comprises the neighbouring municipalities of the Pskov region (Russia) and South-East Estonia.

Research Methodology

The tourism and recreation component often attracts the attention of international experts on euroregion studies (Studzieniecki 2005). Another focus of research is transboundary cooperation in tourism (Prokkola 2008; Timothy 1999; Wachowiak 2006). Besides, numerous studies examine transboundary regions emerging from collaborations in tourism (Vodeb 2010). In Russia, studies in the tourism and recreation component of transboundary regions began with the emergence of the theory of region formation (Baklanov and Ganzey 2004; Fedorov and Korneevets 2010; Korneevets 2010).

In 2010, the theory of transboundary tourism and recreation regions was created based on the earlier findings. The theory was tested using the case of the Baltic Sea region (Kropinova 2011, 2014, 2016). This theory helped to identify the features of the formation of transboundary tourism and recreation regions and to launch pilot case studies focusing on the South-East Baltic. Another mesoregion within the Baltic macroregion that has a strong tourism and recreation component is that bringing together Latvian, Estonian and Russian (Pskov region) municipalities (Pskov region). This mesoregion shares many common features with the Pskov-Livonia euroregion created in 2004.

The Results of the Research

Based on the principle of voluntary participation, the Pskov-Livonia euroregion has ever-shifting boundaries. This fact is reflected in the full name of the region—the Association on cross-border cooperation "Pskov-Livonia". Moreover, the Association does not meet the main criteria for euroregios. I believe that a major obstacle to transboundary regionalisation at the crossroads of Estonia, Latvia, and Russia is the unwillingness of the local authorities (particularly, those on the Russian side) to use European mechanisms to stimulate development. The lack of a common vision of transboundary regionalisation poses another political and institutional obstacle. Furthermore, Estonia and Latvia are more interested in further integration in the European Union than in transboundary regionalisation (Barinov and Kiryushin 2013).

Tourism is a principal area of cooperation in the Pskov-Livonia euroregion. The euroregion's strengths relating to tourism and recreation are as follows: (1) clean environment, free from industrial production and other sources of pollution; (2) cultural and historical monuments—major tourist attractions. However, tourism development is the euroregion is hampered by: (1) a weak economic backdrop; (2) the sparse population (Barinov and Kiryushin 2013, p. 208–209).

On the one hand, recreation and tourism are major elements of the "complex" Pskov-Livonia euroregion. On the other hand, the potential for the development of tourism and recreation varies across the Europregion. It would be a mistake to equate the Pskov-Livonia euroregion with the transboundary tourism and recreation region. In particular, the tourism and recreation region is not based on the principle of voluntary participation.

Despite the economic crisis, the political tensions between Russia, on the one hand, and Estonia and Latvia, on the other, and the accession of the latter countries to the EU in 2004, Russo-Estonian cross-border tourism was booming in the 2000s. Favourite destinations for travellers from the Pskov region, Estonia and Latvia are surpassed only by popular resorts. Moreover, Estonians and Latvians account for most international arrivals at the Pskov region (Turchenko 2014). Driven primarily by shopping, cross-border tourism can nevertheless facilitate the development of the tourism and recreation region.

An analysis of statistics on tourism development in Estonia, Latvia and the Pskov region (Kropinova 2016) shows that this Russian territory does not lag behind its western neighbours. Moreover, if only the areas included in the emerging transboundary tourism and recreation region are considered, the Pskov municipalities receive a greater number of tourists than its Estonian and Latvian counterparts do. This is no surprise since the city of Pskov and the contiguous Pechory district are popular tourist destinations.

Dr. Elena Kropinova suggests considering ten (six primary and four secondary) factors when identifying a transboundary tourism and recreation region (Kropinova 2016, p. 89). What are these factors in the case of the neighbouring Estonian, Latvian and Russian municipalities?

Note that the Russo-Estonian part of the Pskov-Livonia euroregion is in a more advantageous position as regards the development of transboundary tourism and recreation ties than its Russo-Latvian part is. This relates specifically to the process of regionalisation. Therefore, I will focus on the formation of the "Pskov—Pechory —Tartu" tourism and recreation region and identify its unique strengths. An analysis of the factors of region formation makes it possible to carry out the zoning of the euroregion or, in other words, to identify transboundary tourism and recreation microregions of different hierarchic levels within the euroregion.

Three of the four primary factors of region formation operate in both the Russo-Estonian and the Russo-Latvian parts of the euroregion. These are: (1) territorial continuity, i.e. a direct transport link that does not require leaving the territory of the transboundary region; (2) the complementarity of natural, cultural, and historical potentials of local tourism; (3) a common or coordinated transport infrastructure. The three other factors of region formation are less pronounced in the case of the euroregion. These factors are: (4) close ties between tourism actors being a part of the transboundary region; (5) transboundary tourist routes; (6) governmental and/or public institutions responsible for organizing and coordinating transboundary tourist flows.

An assessment of the six factors, using the token approach, shows that all of them operate in different parts of the Pskov-Livonia euroregion. Nevertheless, the Russo-Estonian part of the transboundary region is at the intermediate stage of development, whereas the Russo-Latvian part is still at the early stage. This is explained by the lack of developed tourism and recreation areas either side of the Russo-Latvian border. Although the Pskov region boasts a major tourist attraction —Alexander Pushkin's estate Mikhaylovskoe Manor, which is located close to the Latvian border, it is not highly probable that it will gain transboundary significance. Another tourist attraction, which is situated at the crossroads of Russia, Latvia, and Belarus, is the Sebezh national park. However, it is not easily accessed from the neighbouring countries.

Nevertheless, secondary factors of transboundary region formation seem to be of much greater interest to this research.

The *first secondary factor* of the formation of a tourism and recreation region is the "level of socioeconomic development" (Kropinova 2016). Both the Pskov region and South-East Estonia belong to the respective national periphery. Moreover, household incomes in the Pskov region are among the lowest in the country. In this situation, tourism will boost the economic development of the two countries' periphery.

The *second factor* is the local ethnic identity, i.e. the potential for the development of ethnic and ethnographic tourism. Overall, the Russo-Estonian border marks an ethnic and denominational divide. An exception is the Finno-Ugric minority of Setos living on either side of the border—in the Pechory district of the Pskov region (approximately 250,000 people) and in the Põlva and Võru districts of Estonia. From the 18th century until the 1920s, these lands were part of the Pskov Province. Unlike the Estonians, the Setos profess Eastern Orthodoxy. Having converted in the 16th century, they created a unique culture that combines elements of Christianity and heathenism. The Seto culture has become a special tourist attraction on either side of the border. In Estonia, the Setos have the status of an independent ethnic group. In Russia, they have been included in the list of indigenous ethnic minorities since 2010 (Manakov et al. 2018).

The *third factor* is the common historical past as ground for the development of historical and cultural tourism. Although this secondary factor of region formation does not operate to full capacity here, there is room for the development of historical and cultural tourism. This territory belonged to one state some 250 years (1721–1917 and 1940–1991), whereas the territory of the transboundary region was divided by a national border for over five centuries (from the 13th century to 1721 and from 1920 to 1940) and it is divided today (Manakov and Turchenko 2017).

An exception is Setomaa region. From the 13th century until 1920, it was part of the Pskov Land (later, the Pskov Province). From 1920 to 1944, all of Setomaa belonged to Estonia under the name of Petserimaa (Petseri is the Estonian for Pechory). In 1944, the region was divided between Estonia and Russia. Nevertheless, until 1991, this region belonged to one state only. Thus, the two parts of Setomaa have experienced political isolation for twenty-seven years only (Manakov et al. 2018).

The *fourth secondary factor* is networks of social ties in culture, sport, medicine, research, etc. In the transboundary region, strong social ties have been established between the two cities of Pskov and Tartu, which serves as the regional "cores". This transboundary tourism and recreation region is "double-cored", which translates into an additional regional advantage.

Discussion

The Strategy for the Socioeconomic Development of the Pskov region until 2020 (2009, p. 43) includes a map of local tourist areas. Out of the four areas featured on the map, two are of the first level (Pskov—Pechory and Pushkinskiye Gory) and two are of the second level (Chudskoe Lake and the Sebezh national park). Three of them, which are located at the Estonian and Latvian borers (Chudskoe Lake, Pskov —Pechory and the Sebez national part), lie within the borders of the Pskov-Livonia euroregion.

The above tourist and recreation areas can be viewed as microregions of different levels—Pskov—Pechory as a microregion of the first level and Chudskoe Lake and the Sebezh national park as those of the third level. However, since the Russo-Estonian and Russo-Latvian borders have a strong barrier function, Chudskoe Lake and the Sebezh national park cannot be considered transboundary tourism and recreation microregions proper. This applies to a greater degree to Chudskoe Lake and to a lesser degree to the Russo-Latvian border southeast of the Pskov region. Thus, the mentioned tourist areas are classed as transboundary tourism and recreation microregions at an early stage of development or as potential microregions. To become full-blown transboundary microregions, they require new transboundary transport links and a change in the border control regime (Manakov and Golomidova 2018).

Furthermore, the Pskov—Pechory tourism and recreation microregion stretches into the territory of Estonia with the Setomaa microregion straddling the border. Historically populated by the ethnic minority of Setos, Setomaa is a third-level microregion that covers a considerable part of the Pechory district and the neighbouring Estonian municipalities of Võrumaa and Põlvamaa (Manakov et al. 2018). The two Estonian municipalities and the Pechory district constitute a transboundary microregion of the second level, which can be referred to as "Setomaa and the environs".

Conclusion

Coined in the 2010s, the term "transboundary tourist and recreation region" is new to the Russian research literature. Its emergence followed the development of the theory of transboundary region formation. Internationally, this theory gave rise to the concept and phenomenon of euroregions. A transboundary tourism and recreation mesoregion is the territory at the crossroads of Latvia, Estonia and Russia, where the Pskov-Livonia euroregion was created in 2004. However, it would be erroneous to equate the Pskov-Livonia euroregion with the emerging transboundary tourist and recreation region. Although the prospects for tourism and recreation development vary across the euroregion, all of its parts can contribute to the formation of the tourist and recreation mesoregion.

A tourist and recreation microregion of the third level, Setomaa can be viewed as the geographical "heart" of the transboundary Russo-Estonian region (a microregion of the first level). The microregion has a "core" on either side of the border. These are the cities of Pskov and Tartu. This double-cored region, which can be referred to as "Pskov—Pechory—Tartu", serves as a foundation for the emerging Latvian-Estonian-Russian tourism and recreation mesoregion. Later, having incorporated microregions of a lower level, it may expand north (the Chudskoe Lake microregion) and south-west (Latvian territories). Overall, despite the considerable length of the Russo-Latvian border, the formation of the Latvian part of the mesoregion remains a long-term prospect.

References

Baklanov P, Ganzey S (2004) Prigranichnye i transgranichnye territorii kak ob"ekt geografich-eskikh issledovanii (Border and transboundary territories as an object of geographical research). Proc Russ Acad Sci Ser Geogr 4:27–34 (in Russ.)

Barinov S, Kiryushin P (2013) Kharakter i rezul'taty razvitiya transgranichnykh regionov v razlichnykh institutsional'nykh usloviyakh na primere evroregionov «Eresunn» , «Pskov-livoniya» i «Dnepr» (Nature and results of cross-border regions development in different institutional conditions: the example of the 'Oresund', 'Pskov-Livonia' and 'Dnepr'). Bull Int Organ 8(4):196–213 (in Russ.)

Fedorov G, Korneevets V (2010) Transboundary regions in the hierarchical system of regions: the system approach. Baltic Reg 2:32–41

Korneevets V (2010) Formirovanie transgranichnykh mezoregionov v Pribaltike (Formation of transboundary mesoregions in the Baltic). Immanuel Kant Russian State University, Kaliningrad (in Russ.)

Kropinova E (2011) Factors forming the transboundary tourism and recreation region 'South-Eastern Baltic'. Baltic Reg 1:106–114

Kropinova E (2014) Proekt «Perekrestki 2. 0» v formirovanii transgranichnogo turistskogo regiona Yugo-vostochnoi Baltiki (Project 'Crossroads 2.0' in the creation of the transboundary tourist region in the South-Eastern Baltic area). Pskov Regionological J 17:53–59 (in Russ.)

Kropinova E (2016) Transgranichnye turisticheskie i rekreatsionnye regiony v Baltiiskom more (Transboundary tourism and recreation regions in the Baltic sea). Immanuel Kant Baltic Federal University, Kaliningrad (in Russ.)

Manakov A, Golomidova E (2018) Estimating the development of the Latvian-Estonian-Russian transboundary tourism and recreation region. Baltic Reg 10(1):130–141

Manakov A, Golomidova E, Krastev V, Yancheva K (2018) The development potential of the transboundary tourism and recreation microregion 'Setomaa'. Pskov Regionological J 34 (2):102–116

Manakov A, Turchenko E (2017) Osobennosti formirovaniya transgranichnogo turistsko-rekreatsionnogo regiona «Pskov—Pechory—Tartu» (Features of forming the transboundary tourism and recreation region 'Pskov—Pechory—Tartu'). Pskov Regionological J 32(4):31–39 (in Russ.)

Portal of public authorities of the Pskov region (2009) Strategiya sotsial'no-ekonomicheskogo razvitiya Pskovskoi oblasti do 2020 goda (Strategy for socio-economic development of Pskov region until 2020). http://www.pskov.ru/sites/default/files/documents/2010/soc_econ_strategy_pskov_2020.pdf. Accessed 16 June 2018 (in Russ.)

Prokkola E-K (2008) Resources and barriers in tourism development: cross-border cooperation, regionalisation and destination building at the Finnish-Swedish border. Fennia 186(1):31–46

Studzieniecki T (2005) Euroregions—new potential destinations. Tourism Rev, AIEST 60(4):26–32

Timothy D (1999) Cross-border partnership in tourism resource management: international parks along the US-Canada border. J Sustain Tourism 3–4:182–205

Turchenko E (2014) Geografiya i dinamika mezhdunarodnogo v"ezdnogo i vyezdnogo turizma v Pskovskoi oblasti v 2004-2012 gg. (Geography and dynamics of the international incoming and outcoming tourism in Pskov region in 2004–2012). Pskov Regionological J 18:56–67 (in Russ.)

Vodeb K (2010) Cross-border regions as potential tourist destinations along the Slovene-Croatian frontier. Tourism Hospitality Manag 16(2):219–228

Wachowiak H (2006) Tourism and borders. VT Ashgate, England

The Universities and Economic Modernisation in the Bordering Regions of Russia and Belarus

Alexander P. Katrovskiy, Tatyana I. Pototskaya and Svetlana A. Sergutina

Abstract

This article examines the correlation between economic development and higher education in the bordering regions of Russia and Belarus. We stress that, in post-industrial society, universities are becoming a major driver of innovative regional socio-economic development and modernisation. This study aims to establish how universities affect the economic development of the bordering regions of Russia and Belarus. The development of universities and engagement in higher education are important indicators of regional competitiveness in the knowledge economy. We emphasise the need to create universities that are capable of promoting regional innovative development. The top-ranking Russian universities and the Belarusian border calls for the establishment of the so-called "universities promoting regional innovation". Innovative regional development requires synergy between higher education and the economy. Economic modernisation must precede the creation of new universities. Otherwise, graduates have no other choice than to move to more developed regions. The Bryansk and the Pskov regions rank low in the education index, which has an adverse effect on the local high-tech industries. This research is the first attempt to compare the bordering regions of Russia and Belarus in terms of university graduates in total population and engagement in higher education.

Keywords

Bordering regions of russia and belarus · Higher education · Regional modernisation · Educational background

A. P. Katrovskiy (✉) · T. I. Pototskaya
Smolensk State University, Smolensk, Russia
e-mail: alexkatrovsky@mail.ru

S. A. Sergutina
Smolensk Institute of Economics, Smolensk, Russia

© Springer Nature Switzerland AG 2020
G. Fedorov et al. (eds.), *Baltic Region—The Region of Cooperation*,
Springer Proceedings in Earth and Environmental Sciences,
https://doi.org/10.1007/978-3-030-14519-4_6

Introduction

Educational background is a basic characteristic of the human capital of the region. Moreover, it is an important indicator of the regional investment climate and innovative development. The proportion of university graduates is a key measure of a population's education. The educational background of the regional population is strongly affected by the situation in, and the development of the regional educational system. As a rule, regions with more developed higher education systems have a higher proportion of university graduates. Another important measure is engagement in higher education, i.e. the number of students per 10,000 population. As to regional and cross-country comparisons, the Education Index has been incorporated in the Human Development Index (Lisichenok and Chernyavskii 2013; Bobylev and Grigor'ev 2016).

Differentiation in education—differentiation in development. More than 120 years ago, the famous Russian economist and educationalist, Professor of Moscow Imperial University, Ivan Ivanovich Yanzhul, wrote about the influence of education on national development: "Once Russia is educated, it will be rich too" (Yanzhul 1896). This phrase has a bearing on the current discussion about the trajectories of economic development in the bordering regions of Russia and Belarus. Economic modernisation is impossible without innovative development, which in its turn requires significant investment in education, particularly higher education. Many researchers have stressed this fact (Emel'yanov and Khachaturyan 2011; Koritskii 2010). However, doubts have been expressed as to the demand for human capital in today's Russia (Gimpelson 2016). The problem of human capital development in the bordering regions of Russia and Belarus has been discussed in a monograph by Russian and Belarusian geographers (Katrovskii and Kovalev 2017). Innovative regional development requires a favourable climate for innovations. Thorsten Hagerstrand stresses that the efficacy of innovation diffusion depends on the retransmission capacity of cities and regions to a much greater degree than on the distance, the former being strongly affected by the human capital (Hagerstrand 1967).

As society is trading the path of post-industrial development, the measures of the population's education (including the proportion of university graduates) are becoming increasingly important indicators of regional development. An analysis of census data shows that the number of university graduates has significantly increased with the bordering regions of Russia and Belarus over the past sixty years. However, in 1959 and 2009–2010, the number of university graduates was below the respective national averages both in the Russian and Belarusian regions. The gap between these regions and the respective national capitals was even wider (Table 1).

In the Soviet period, a major trend was the growing proportion of educated people. The literacy rate and the proportion of university graduates traditionally are the two traditional measures of a population's education.

An important trend observed in the bordering regions of Russia and Belarus in the post-Soviet period is the narrowing of the gap between these regions and the national averages. In 1989, the education index in the Gomel region (the most developed area of the Belarusian borderlands) was 18% below the average performance of the Byelorussian Soviet Socialist Republic. Twenty years later, the

Table 1 The changes in the proportion of university graduates in the two countries and their regions, %

	1959	1989	2002 for RF, 1999 for RB	2010 for RF, 2009 for RB
Russian Federation (RF)	1.93	8.64	13.12	19.28
Moscow	9.02	21.72	26.56	36.24
Bryansk region	0.97	6.05	10.27	15.75
Smolensk region	1.18	6.84	11.74	16.67
Pskov region	0.97	6.48	10.93	15.67
Republic of Belarus (RB)	1.19	8.31	11.28	16.10
Vitebsk region	0.94	6.81	9.62	13.87
Gomel region	1.00	6.83	9.00	13.58
Mogilev region	1.03	6.66	8.99	13.31

Source RSAE (1959, 1989), National Statistical Committee of the Republic of Belarus (2011)

performance of the Vitebsk region (the most advanced border territory at the time) was only 14% below the national average. Even greater "convergence" is being observed in the Russian Federation. In 1989, the performance of the Smolensk region in terms of education was 21% below the average across the Russian Soviet Federative Socialist Republic. In 2010, the region was only 14% below the national average. The gap between the Russian regions and Moscow narrowed sharply. In 2010, the Smolensk region had 2.17 times fewer universities, the Bryansk region 2.3 times, and the Pskov region 2.32 times graduates than Moscow. In 1989, these figures were 3.18, 3.59, and 3.35 times respectively (Katrovskii 2013).

The proportion of university graduates in different bordering regions of Russia and Belarus depends both on the level of local universities' development and the regional socio-economic development. A key measure of university development is engagement in higher education, i.e. the proportion of students in total population. Since this is a relative measure, it ensures a reliable comparison of regions with different population numbers (Table 2).

The academic year 2015/16 witnessed a significant reduction in both the absolute and relative measures of engagement in higher education across the bordering regions of Russia and Belarus. The number of students per 10,000 people was only 69.7% of the level of 2010/11 in the Smolensk region, 64.4% in the Pskov region, and 72.2% in the Bryansk region. A slightly less dramatic decrease was observed in Belarus (67.3% in the Gomel region, 75.8% in the Mogilev region, and 89.2% in the Vitebsk region). In 2016, the Belarusian border regions (285) outperformed the Russian border regions in terms of engagement in higher education. The best performing areas are the Vitebsk and Mogilev regions, whereas the Pskov and the Gomel are ranked at the bottom. At the same time, the annual decrease in the rate of engagement in higher education by 8–12% over the past five years can be explained by the rapid development of higher education in the previous years rather than by a crisis or a loss of interest in obtaining an education.

All the Russian regions have improved their position in the Education Index in the past ten years. However, in 2014 (when this indicator was last calculated), the Bryansk (0.923) and the Pskov (0.923) regions deviated from the national average

Table 2 The changes in the proportion of students in total population in the two countries and their regions, 2005–2015

	The number of students per 10,000 population			
Academic year	2005/06	2010/11	2012/13	2015/16
Russian Federation	493	497	424	325
Central Federal District	569	597	479	372
Bryansk region	345	443	387	282
Smolensk region	404	499	424	273
Moscow	1097	1106	786	616
Northwestern Federal District	534	530	451	335
Pskov region	324	382	327	222
Kaliningrad region	428	446	398	280
Saint Petersburg	910	935	759	580
Republic of Belarus	398	467	453	354
Vitebsk region	261	342	375	305
Gomel region	352	394	373	265
Mogilev region	358	377	365	286

Sources National Statistical Committee of the Republic of Belarus (2017), Federal State Statistics Service (2016a)

(0.933) and the average across the Central and the Northwestern Federal Districts. The gap between the performance of these two regions and that of Moscow, St. Petersburg, and the Kursk and the Orel regions was even more considerable.

The bordering regions of Russia and Belarus need a new policy on education and innovations. Projects aimed at developing and putting into practice the latest research advancements in various sectors of the economy must receive public support. This, in turn, requires the restructuring of higher education and the creation of universities capable of providing high-quality education. The absence of top-ranking universities in the border regions precludes innovative development.

In 2016, the proportion of innovative companies was rather low in the Russian regions bordering on Belarus. This proportion was as low as 5.9% in the Smolensk region, 6.7% in the Bryansk region and 8.0% in the Pskov region. The national average was 8.8%, whereas the proportion of innovative companies reached 9.8% in the Central and 8.9% in the Northwestern federal districts. In Moscow, this proportion was 18.0%, in St. Petersburg 16.8% (Federal State Statistics Service 2016b)

This disparity is largely due to the poor development of higher education in the Russian regions bordering on Belarus. At the same time, the improvement of the higher education systems in the bordering regions of Russia and Belarus cannot be regarded as the ultimate goal. The modernisation of higher and secondary vocational education should take into account the current and projected sectoral and spatial structure of the economy. Moreover, modernisation should be accompanied by the emergence of new local high-tech companies, which will create jobs for the graduates. Otherwise, the graduates will have to move to other regions where labour markets offer better employment opportunities for highly skilled professionals. The

Russian border region may adopt the Smart Specialisation approach, which was developed as part of the Europe 2020 strategy for innovative development across the European Union. This approach suggests the simultaneous development of higher education, high-tech industries and the regions' unique innovation sector (Aralica and Bačić 2017).

Higher education is an important driver of regional development. Thus, it is necessary to pursue a policy aimed at supporting regional universities. Moreover, support for regional higher education institutions may help to reduce the inequality in living standards, to prevent the metropolitan universities from draining the periphery of the best applicants, and to provide Russian regions with skilled labour. Regional support makes it possible for the local universities to become major research centres and drivers of innovative development. This has taken place since 2016.

Since most Russian regions specialise in industrial production, agriculture, or services, the regional universities should train specialists in the relevant fields.

The universities focusing on the training of teachers, economists, or lawyers cannot be granted the status of "universities promoting regional innovation" (this status and associated funding have been given to Russian higher education institutions since 2016). However, these universities can become the core around which regional universities will grow by merging with technical universities or their branches. A "university promoting regional innovation" should ensure the innovative development of the regional economy in the present and the future. The desired result is not a mechanical merger of several universities but rather the careful selection of study programmes and research project. This is especially true for the bordering regions of Russia and Belarus where the universities that evolved from pedagogical institutes serve as centres for sciences, the humanities, research, and public education.

The problem of innovative development of the Russian regions bordering Belarus is exacerbated by the fact that the metropolitan national research universities, charged with developing technological breakthroughs increasing the competitiveness of the Russian economy, are geographically remote from the main regions where the graduates will probably seek employment. One cannot expect the graduates of metropolitan universities to move to the border regions. Moreover, the cleavage between the centre and the periphery also complicates the economic modernisation of the border regions. Highly developed capital cities attract the most qualified human resources from the periphery. In 2015, per capita income in the Smolensk region was estimated at 40% of that in Moscow and at 63% of that in Saint Petersburg. These proportions reached 40 and 63% in the Bryansk region and 36 and 56% in the Pskov region (Federal State Statistics Service 2016a). Against this background, most graduates of the regional universities plan to move to the metropolitan regions. The loss of human capital to out-migration necessitates both the dramatic modernisation of the entire system of education and training and profound changes in the regional social policy. To an extent, one can agree with Richard Florida who wrote: "Government has its most important and legitimate role to play in establishing the enabling framework for a new era of shared prosperity, and it squanders precious resources that could support such future-oriented, prosperity-boosting efforts when it chooses to bail out old industries, breath life back into outmoded institutions, or place Band-Aids on problems". (Florida 2012)

Conclusion

Today, the bordering regions of Russia and Belarus lag behind their counterparts in education, which complicates innovative development, poses an obstacle to the overcoming of the periphery states, and has a negative effect on the investment climate. The economy modernisation of the bordering regions of Russia and Belarus is impossible without profound changes in the education system.

One of the architects of the Emancipation Reform of 1861, Yakov A. Soloviev, contemplated the prospects and problems of the transformation of the Smolensk Province's economy in his article "The Present and the Future of the Smolensk Province", which was published 160 years. In particular, he wrote: "What do we want for the Smolensk Province? The same thing as for all of Russia—railways, schools and the right public opinion: everything else will come by itself" (Soloviev 1857). Just like 160 years ago, successful modernisation of the bordering regions of Russia and Belarus requires a better infrastructure (particularly, the upgrading of the transport system), top-ranking universities, and institutional transformations. Only through such changes will the bordering regions of Russia and Belarus enter the ranks of the most developed regions of the Eurasian Economic Union.

References

Aralica Z, Bačić K (2017) Regional competitiveness in the context of "New industrial policy"—the case of Croatia. Zb Rad Ekon Fak Rij 35(2):551–582

Bobylev SN, Grigor'ev LM (eds) (2016) Tseli ustoichivogo razvitiya OON i Rossiya. Doklad o chelovecheskom razvitii v Rossiiskoi Federatsii (The goals of sustainable development of the United Nations and Russia. Report on Human Development in the Russian Federation). ac.gov. ru/files/publication/a/11068.pdf. Accessed 7 Aug 2017 (in Russ.)

Emel'yanov Y, Khachaturyan AA (2011) Chelovecheskii kapital v modernizatsii Rossii: institutsional'nyi i korporativnyi aspekty (Human Capital in Russia's Modernisation: Institutional and Corporate Aspects). Moscow (in Russ.)

Federal State Statistics Service (2016a) Regiony Rossii. Sotsial'no-ekonomicheskie pokazateli 2016. Stat. sbornik (Regions of Russia. Socio-economic indicators 2016. Stat. collection). Moscow (in Russ.)

Federal State Statistics Service (2016b) Udel'nyi ves organizatsii, osushchestvlyayushchikh tekhnologicheskie innovatsii v obshchem chisle obsledovannykh organizatsii (po sostoyaniyu na 29 aprelya 2016 goda) (The proportion of organizations implementing technological innovations in the total number of organizations surveyed (as of April 29, 2016). http://www. gks.ru/free_doc/new_site/rosstat/pok-monitor/pok-monitor.html. Accessed 15 Dec 2016 (in Russ.)

Florida P (2012) Bol'shaya perezagruzka. Kak krizis izmenit nash obraz zhizni i rynok truda (Great reboot. How the crisis will change our way of life and the labor market). Moscow (in Russ.)

Gimpelson VE (2016) Nuzhen li rossiiskoi ekonomike chelovecheskii kapital? Desyat' somnenii (Does the Russian economy need human capital? Ten doubts). Vop Ekonomiki 10:129–143 (in Russ.)

Hagerstrand T (1967) Innovation diffusion as a spatial process. University of Chicago, Chicago and London

Katrovskii AP (2013) Transformatsiya vysshego obrazovaniya na postsovetskom prostranstve: ekonomiko-geograficheskie aspekty izucheniyam (Transformation of higher education in the post-Soviet space: economic and geographical aspects of the study). Regional'nye Issledovaniya 4:19–31 (in Russ.)

Katrovskii AP, Kovalev Y (2017) Chelovecheskii kapital i sotsial'no-ekonomicheskoe razvitie regionov rossiisko-belorusskogo prigranich'ya (Human capital and socio-economic development of the regions of the Russo-Belarusian borderland), vol 2. Universum, Smolensk (in Russ.)

Koritskii AV (2010) Chelovecheskii kapital kak faktor ekonomicheskogo rosta regionov Rossii (Human capital as a factor of economic growth in Russian regions). Novosibirsk (in Russ.)

Lisichenok SI, Chernyavskii Y (2013) Vozmozhnosti primeneniya IRChP dlya sravnitel'nogo analiza regional'nogo razvitiya. Osnovnye pokazateli IRChP Respubliki Belarus' za 2012 g. (The possibilities of using the HDI for a comparative analysis of regional development. The main indicators of the HDI of the Republic of Belarus for 2012). https://42.tut.by/331413. Accessed 7 Jan 2017 (in Russ.)

National Statistical Committee of the Republic of Belarus (2011) Perepis' naseleniya 2009. Obrazovatel'nyi uroven' naseleniya Respubliki Belarus' (Population census 2009. Educational background of the population of the Republic of Belarus), vol 4. Minsk (in Russ.)

National Statistical Committee of the Republic of Belarus (2017) Obrazovanie v Respublike Belarus': statisticheskii sbornik (Education in the republic of Belarus: statistical collection). http://www.belstat.gov.by/ofitsialnaya-statistika/solialnaya-sfera/obrazovanie/publikatsii_8/index_7499/. Accessed 17 Oct 2017 (in Russ.)

RSAE (1959) Vsesoyuznaya perepis' naseleniya 1959 goda. Tablitsa 7. Raspredelenie naseleniya po vozrastu i urovnyu obrazovaniya (All-union population census of 1959. Table 7. Population distribution by age and educational background). http://www.demoscope.ru/weekly/ssp/rus_edu_59.php. Accessed 26 Dec 2016 (in Russ.)

RSAE (1989) Vsesoyuznaya perepis' naseleniya 1989 g. Tom 6. Tablitsa 2. Raspredelenie naseleniya SSSR, soyuznykh i avtonomnykh respublik, avtonomnykh oblastei i okrugov, kraev i oblastei po urovnyu obrazovaniya i vozrastu (All-union population census, 1989 Volume 6. Table 2. Distribution of the population of the USSR, union and autonomous republics, autonomous regions and districts, territories and regions in terms of educational background and age). http://www.demoscope.ru/weekly/ssp/rus_edu_89.php?reg=2. Accessed 26 Dec 2016 (in Russ.)

Soloviev Y (1857) Nastoyashchee i budushchee Smolenskoi gubernii (Present and future of the Smolensk province). Ekonomicheskii Ukazatel' 11:258–262 (in Russ.)

Yanzhul II (1896) Znachenie obrazovaniya dlya uspekhov promyshlennosti i torgovli (The importance of education for success in industry and trade). Tekhnicheskoe obrazovanie 3 (in Russ.)

Russia and the Baltic States: Some Results Interstate Relations

Nikolai M. Mezhevich⊙ and Natalia Y. Markushina⊙

Abstract

Geographically, the Baltic States border on Russia, and, in this sense, both parties objectively need to build constructive relations with each other on a range of issues. However, over the sovereign period after collapse of the USSR numerous thorny problems have been plaguing the relationships between the countries concerned. Arguably, in the contemporary span of time an array of dramatic discrepancies arises between Russia and the Baltic states. They should be sorted out and this settlement is vital for stable social-economic development of the entire Baltic region. Still, explicit anti-Russian policy and sentiments of the Baltic elites impede prospects for development of the mutually advantageous cooperation. Noteworthy, in an effort to design a new model of relationships between Russia and the Baltic States security issues along with the interests of the European Union (EU) and North-Atlantic Treaty Organization (NATO) feature a stumbling block. The attempts at exploiting historical memory also are another specific trait of the existent model. All these factors raise the question of accountability of the Baltic countries for the conducted policy and acceptance of its implications. From the authors' viewpoint, it is a dead-end model. The aim of this research is to find out specific characteristics of interaction between Russia and the Baltic States in the contemporary period of time. The interdisciplinary methodological scientific approaches, as well as the multifactorial balance methodology have been used, which make it possible to analyze dynamics and continuity of interstate relations within wide historical frameworks, since it is

N. M. Mezhevich (✉)
Faculty of International Relations, Department of European Studies,
St. Petersburg State University, St. Petersburg, Russia
e-mail: mez13@mail.ru

N. Y. Markushina
Faculty of International Relations, Department of World Politics,
St. Petersburg State University, St. Petersburg, Russia
e-mail: N.markushina@spbu.ru

© Springer Nature Switzerland AG 2020
G. Fedorov et al. (eds.), *Baltic Region—The Region of Cooperation*,
Springer Proceedings in Earth and Environmental Sciences,
https://doi.org/10.1007/978-3-030-14519-4_7

53

impossible to conceive of the modern politics without the entire genesis context. The conclusion was made: despite awareness of benefits from economic cooperation, the political and ideological nuances annihilate an opportunity for a dialogue. Forecasts for evolution of bilateral relations are quite pessimistic. Yet, further exacerbation of the Russian-Baltic relations is capable of confounding the internal policy of the Baltic States as well.

Keywords

The baltic region · The baltic states · Foreign policy · Russia's foreign policy · Political elites · Foreign economic affairs · Historical memory

The Baltic Sea region remains a forefront dimension in foreign policy of the Russian state. Having given an insight to history, this vector has been steady, in spite of the tricky political environment, which had been accompanying nascence of the Russian statehood. The three points can be figured out:

1. In N. M. Mezhevich's opinion, "historical value of the Baltic and Scandinavian directions within the policy of the Russian state has been determined by uninterrupted continuity of the Russian statehood in that direction, even in the feudal dispersion era and Tatar-Mongol invasion" (Mezhevich 2004).
2. Economic bonds, as a crucial element in the Baltic vector of the Russian state, have always been taking precedence over the political ambitions.
3. And, finally, a vast amount of actors on quite a small geographic territory has been the third constituent in the nascent relations model.

Nevertheless, after collapse of the USSR for a long period of time Russia had not had common understanding, which foreign policy with regards to the Baltic States it was to opt for. In 1990s there was a hope for maintenance of ties and pragmatic contacts, but there was no relevant strategy. The desire to establish relationships with "Grand Europe" under the Foreign Minister A. Kozyrev implied increased attention to Berlin, Paris, London, and Brussels—the traditional benchmarks in even the pre-Soviet policy.

That situation began changing, when "Grand Europe" itself turned to amending the game rules, having "securitized" politics in the Baltic Sea region.

Therefore, nowadays, we can indicate a series of factors, which are affecting the modern relations model within the Baltic region:

1. The Baltic States is a neighborly region, social-economic processes whereof might bring about either positive, or negative influence on adjacent Russia.
2. Having acceded to the EU and NATO, Estonia, Latvia and Lithuania feature a unique case of the post-Soviet states. The outcomes of their political-economic development are meaningful to such countries as Ukraine, Georgia, Moldova, drifting the same direction and other member-states of the Eastern Partnership, including Belorussia.

3. The Baltic States are absolute leaders in costs per capita on military programmes within NATO. On their territory NATO military exercises, involving Finland and Sweden, have been going on almost continually over the recent three consecutive years.
4. The Baltic States are the most zealous supporters, and at times the chief ideologists of Russo-phobic attacks at Russia, the anti-Russian foothold in the post-Soviet space. Interestingly, such a role is caused by a certain economic substantiation and, equally important, economic implications.
5. The Baltic States is a transit rival to Russia, however, under definite circumstances the Baltic States region has proved to be a bridge between the West and Moscow.
6. Acute maturity of interregional ties is also remarkable. The North-Western Federal District has been actively developing cooperation between subjects of the Russian Federation and the Baltic States.

Still, if we are going to devise a new model of relationships between Russia and the Baltic States, our reflections should stem from an array of points.

1. *Shifts in the political landscape.*

The new starts of electoral cycles (parliamentary elections in Latvia in October 2018 and in Estonia in March 2019, presidential elections in Lithuania in May 2019), which were held in the later half of 2018 and early half of 2019 cannot but shake the existing political and social-economic development patterns of the Baltic States.

Russo-phobia of their political elites has practically run out of its financial-economic steam. The EU dotation into the Baltic republics after 2020 is to plunge steeply. Economic benefits from NATO troop's deployments in the region are far too hazing. Moreover, it diminishes investment attractiveness.

Demographic situation is catastrophically deteriorating, the population level is declining, the best part of workforce, first and foremost, the youth, are moving to the West. Rampant corruption among the Baltic elites is getting ever more evident. In Estonia this is called "seemukapitalism", i.e. "barons' capitalism". Abidance by the formal rules governing functioning of economic institutions in case of breaking the principles of a real competitive market is another hallmark of economic models prevailing in Estonia, Latvia, and Lithuania.

Prospects for growth of the economy's real sector in the region, including the transit sector, in the context of sanctions warfare with Russia seem to be quite illusive. The Baltic region has turned into the backyards of the EU and the frontline district from the "storefront" of the Soviet Union. Inertial movement towards the same direction is likely to put the issue of economic viability/financial solvency to the agenda of the next decade (after 2021).

Under these circumstances, the electoral cycle 2018–2019 is theoretically able to cause alternations in the political generations—from Russophobes to pragmatists. The latter category denotes political persons, who prefer social-economic benefit to the belligerent rhetoric. Nevertheless, the reverse scenario should not be considered a lost cause: total ethnic mobilization with a view to put the blame on the ethnic minorities and Russia for all hardships piled up over the recent years.

2. *Security issues*

The Ukrainian crisis, "Skripal affair", sanctions policy and ambassadors expulsion, i.e. new elements of the odious Western policy have adjusted the roles of the Baltic countries in the political arena. The new security concept adopted by the Baltic States relies upon the two pillars: absolute confidence that the danger emanates from Russia and irrational policy towards its own citizens.

As such, on March 3rd, 2018 presidents of Latvia (Raimonds Vējonis), Lithuania (Dalia Grybauskaitė) and Estonia (Kersti Kaljulaid) conducted negotiations with president of the USA, Donald Trump. At the summit dedicated to centenary of separation of the Baltic States from Russia, heads of republics asked Trump to reinforce NATO military commitment in the Baltic region and touched upon the issue of the US natural gas supplies (Bovdunov 2018). "The USA expressed readiness to hand over 170 million US dollars to the Baltic countries as a military aid. Another 3 million US dollars is to be allocated for informational resistance, as it was stated in the White House press-release" (Bovdunov 2018). On April 9th, 2018 in Paris a meeting of the Baltic presidents with president of France Emmanuelle Macron was held. There, as press-releases reported, "Great attention should be paid to tackling challenges in the sphere of international security and cooperation in defense sphere" (Macron will of the presidents of the Baltic States in Paris 2018). Undoubtedly, visits to the USA and Europe are essential to the political elites of the Baltic States. The ultimate question—what will they bring about? A range of analysts believe, "being aware of their "consumables" status, Estonia, Latvia and Lithuania solicit the Western allies to concentrate in the Baltic region as many troops as possible, so as to somehow inflate their reputation" (Ishchenko 2018). The fact that Russia is not going to occupy the Baltic neighbors is likely to cause political failure of this policy. But by that time already weak budget would have been shattered by defense expenditures, the promised financial injections would have appeared to be less than it was pledged, or spent on some projects, and countries themselves would have been crowded with Western troops. The migration issue is possibly to arise thornily again. "The experts reckoned: if all emigrants did not return, and emigration kept on, in a couple of years the Baltic States would turn into the retirees' country (mainly those, who would be beyond their sixties)" (The youth went to work in Europe: why the Baltic States are empty the farm and the city 2018). In a similar vein, another fact is notable, Eurostat showed to Latvia residents all-out weakness of social welfare funding in their country. Latvia lags behind practically all European countries, emulating only Romania, with social welfare indicating even lower—14.6% of GDP" (Eurostat: Latvia is the penultimate country in terms of social protection costs 2017). In Lithuania doctors continually complain of non-payment of their wages and take to streets on demonstration almost on a constant basis (Eurostat: Latvia is the penultimate country in terms of social protection costs 2017). And Estonia's report by the State Control Department "presented that the state should prepare for a situation, whereby European compensations and allowances would shrink already in the new budgetary period—according to the figures provided by the Finance Ministry let alone a Brexit impact, approximately by 40%, i.e. by 1.5 billion Euro" (Estonia is preparing to tighten the belt: EU funding will decrease 2017). Yet, Lithuania is ready to invest nearly 770 million euro (the country's GDP is 39 billion euros) in national defense in 2018. Latvia's military budget in 2018 stood at 576 million euro that was 126 million larger than in the previous year (the republic's GDP—about 40 billion euro). Estonia's military budget is approximately 418 million euro (GDP exceeds 20 billion euro)" (Hrolenko 2017). Amid these negative economic indices military strife appears to be quite irrational.

Today the attitude of the Baltic States to Russia is utterly politicized, and the Russian threat is "securitized". What does it usher in? The organic path of relations development implies breakdown of economic ties, lost opportunities in the economy and politics. It can be stated that the adverse economic effect in view of the Russian immensity is almost impossible to calculate, however the Baltic economies suffer from annual losses amounting to 10% of GDP. Political losses are harder to estimate in quantifiable terms. The anti-Russian rhetoric of the Baltic politicians puts certain pressure on the European power elites and public opinion. However, the principal effect is achieved through discrediting Russia as an economic partner, deliberate placing the country's exports at competitive disadvantage. Thereby, we can argue about increasing economic, political, and social risks in the region.

3. *The effect of the European Union in the Baltic region*

In opinion of the expert I. Busygina, "Integration has been acquiring favorable reputation in the eyes of separate countries as they gained specific benefits from implementing specific, more modest cooperation projects along definite cooperation realms, rather than generalist ones" (Busygina 2013).

The matters of integration and cooperation between the EU and Russia within the Baltic region, search for models, which would make it possible to pull out of the worrying situation is still a priority for the Russian scholars. The key issue is, certainly, the economy, which edges us closer to a vision that the previous models of international relations, including international economic relations, are unlikely to endure any more in the nearest future.

In the context of the Baltic region, we are passing over to the situation of norms lack. All parties concerned should work, ensure economic growth, and create their own integration fields. One of the probable variants is to follow the integrational suit of the Eurasian Union with the European Union. But so far, we have accumulated such a volume of discrepancies that it is unclear, how to escape this trap. Too many ideological complexes obfuscate this riddle.

In this sense, a new factor of China's presence, which is interested equally in Russia and states, surrounding it in the post-Soviet area, as well as in Europe, as it is clear that all projects under the aegis of the "new Silk Road" do not mean Moscow as the final station, they include many other stations, with most of them located in Europe. And here, absence of complexes and an aggregate of the Estonian-Chinese, Chinese-Lithuanian, etc. conflicts is likely to play into hands. Building a strategic partnership between Russia and China might gender some kind of an interim format, a new model of new Eurasian integration, particularly given that China has been taking a strong interest in Northern Europe and cooperation with Belorussia.

As far as grand Eurasian integration is concerned, several equal partners are implied. Kazakhstan is another eminent player, which is interested in paving the common corridor, rather than designing routes only to Astana. Such logic might speed up creation of a new relations model. Moreover, we are witnessing that nowadays the European integration is going through not the best of times. It became unsuccessful, when the goal of integration turned into a process as a result of some compound evolution, complicated by vague techniques of spatial expansion combined with innate increasing complexity and bureaucratic immobilization. The era of the European integration had set out after World Wars and came to an end after the Maastricht agreements, when intake of new states triggered almost an unmanageable situation, failed the European Constitution and brought on Brexit. Undoubtedly, the very ideas of the European civilization and European values are eternal; they are still landmarks for Russia. However, the great European idea should not be replaced by expedience of the current political situation, what the Baltic countries are aspiring to.

To conclude, we can state that the position of the Baltic States within the Russian system of foreign political and foreign economic priorities remain crucial. Still, in the recent years analysis of all versions of Russia's Foreign Policy Concept has been testifying that the Baltic countries are becoming less meaningful. Russia, in its turn, is also unwilling to go along for the sake of an advantageous dialogue even at heavy concessions. Obviously, Russia has got tired of practically unilateral attempts at inspiring the dialogue. Theoretically, after shifts in Baltic political elites countries might turn around from the mono-vector foreign policy to its multi-pronged model—i.e. resort to the potential of the traditional geographic-historical ties. However, after 2014 developments such a scenario seems to be almost impossible. A strict anti-Russian course, which has been consecutively pursued in the recent years, is blended in shaping public opinion, political slogans, political, particularly, economic practices. The compatriots issue in the Baltic States features a separate topic for the discussion.

The lesson here is that under the present conditions the Russian policy should not respond to slight changes in foreign policies of our Baltic neighbors. Individual moves by our partners aimed at revision of the course and solution of the most urgent issues should also be taken into account. In particular, Saint-Petersburg, the Leningrad region, the Kaliningrad region are the key agents of the state foreign policy in the region. Development of cooperation within the regional organizations (the Council of the Baltic Sea States, the Union of the Baltic Cities, the Baltic Development Forum) is also prospective. Russia endeavors a lot in conversion of these international organizations into the platforms for joint coordination of foreign affairs within the region.

Russia should garner general amendments of the political course in neighboring countries to the level of pragmatic cooperation, without making trade-offs on those principled stances in politics and economy, which are outlined in this paper.

References

Bovdunov A (2018) Na poklon v Vashington: chego zhdut lidery pribaltijskih stran ot vizita v SSHA (Bow down to Washington: why wait for the leaders of the Baltic countries from the visit in the USA). https://russian.rt.com/world/article/499595-tramp-pribaltika-prezidenty-vstrecha. Accessed 12 June 2019 (in Russ.)

Busygina I (2013) Assimetrichnaya integraciya v Evrosoyuze (Asymmetric integration in the European Union). http://www.intertrends.ru/fifteen/002.htm. Accessed 16 May 2018 (in Russ.)

Eurostat: po raskhodam na social'nuyu zashchitu Latviya zanimaet predposlednee mesto v ES (2017) (Eurostat: Latvia is the penultimate country in terms of social protection costs). http://rus.delfi.lv/news/daily/latvia/eurostat-po-rashodam-na-socialnuyu-zaschitu-latviya-zanimaet-predposlednee-mesto-v-es.d?id=49529229. Accessed 12 June 2019 (in Russ.)

Hrolenko A (2017) Voennye byudzhety stran Baltii dostigli dvuhprocentnogo katarsisa (The military budgets of the Baltic countries made up two percent of catharsis). http://ru.sputniknews.lt/columnists/20171214/4626960/voennye-budzhety-stran-baltii-dostigli-dvuhprocentnogo-katarsisa.html. Accessed 12 June 2019 (in Russ.)

Makron primet prezidentov stran Baltii v Parizhe (2018) (Macron will of the presidents of the Baltic States in Paris). https://www.rubaltic.ru/news/06042018-makron-primet-prezidentov-stran-baltii-v-parizhe/. Accessed 12 June 2019 (in Russ.)

Mezhevich NM (2004) Baltijskij region i Rossiya na Baltike: specifika pozicionirovaniya (Baltic region and Russia on the Baltic sea: specificity of positioning). https://www.ut.ee/ABVKeskus/sisu/publikatsioonid/2004/pdf/VF-B.pdf. Accessed 12 June 2019 (in Russ.)

Molodezh' edet na zarabotki v Evropu: pochemu v Pribaltike pusteyut hutora i goroda (2018) (The youth went to work in Europe: why the Baltic States are empty the farm and the city). http://www.ntv.ru/novosti/2005082/. Accessed 12 June 2019 (in Russ.)

Ishchenko R (2018) Strany Baltii prosyat Zapad ob okkupacii (The Baltic countries asked the West about the occupation). https://novorosinform.org/716398. Accessed 12 June 2019 (in Russ.)

V Estonii gotovyatsya zatyagivat' poyasa: finansirovanie ES umen'shitsya (2017) (Estonia is preparing to tighten the belt: EU funding will decrease) https://eadaily.com/ru/news/2017/12/08/v-estonii-gotovyatsya-zatyagivat-poyasa-finansirovanie-es-umenshitsya. Accessed 12 June 2019 (in Russ.)

Investment and Production Cooperation Between the Countries in the Baltic Region: Current State and Problems

Vladimir I. Chasovsky ⓘ

Abstract

In the article the author tried to focus on one of the key aspects of world economic development—the emerging regional multipolarity. The fact of modern progressive growth of Russian economy is undeniable; therefore, the author sees particular interest in the method of analysis of the place and role of European countries, transnational companies and national firms in the Baltic region in the sphere of investment and production cooperation with each other and with Russia. The scientific novelty of the work is determined by the author's approach and the author's method of research based on the combined use of comparative geographic, statistical, mathematical and historical-geographical methods, which make it possible to determine the current sectoral interests and preferences of the countries of the region, changes in the attitude of the different EU countries to each other, to Russia and various subjects of the regional economy. The analysis of the data presented in the article clearly indicates still insufficient use of Russia's potential in the field of cross-border cooperation and in strengthening the country's economic position in the Baltic region. In addition, it is now understood that the activation of cooperation between the EU and Russia in high-tech production spheres should be the main element of the interaction within the framework of investment and production and other integration processes in the Baltic.

Keywords

Baltic region · European union · Russian Federation · Industry · International cooperation · Cross-border cooperation · Internal structure of the region · Development of economic cooperation

V. I. Chasovsky (✉)
Institute of Environmental Management, Urban Development and Spatial Planning,
Immanuel Kant Baltic Federal University, Kaliningrad, Russia
e-mail: prof.chasovsky@mail.ru

© Springer Nature Switzerland AG 2020
G. Fedorov et al. (eds.), *Baltic Region—The Region of Cooperation*,
Springer Proceedings in Earth and Environmental Sciences,
https://doi.org/10.1007/978-3-030-14519-4_8

Differentiation of Countries and Geographical Segments of Investment and Production Cooperation

Under modern geopolitical and economic conditions, the Baltic region as a zone of differentiation of national interests and simultaneously as a zone of cross-border cooperation attracts attention of both the international community and the scientific community. There is even more reason for this as nowadays the innovative and investment-and-production integration is declared as a priority area of economic cross-border cooperation between the countries of the Baltic region. At the same time, for Russia the development of technological and innovative cooperation with partners in the region is especially urgent due to the increased tensions in relations with Western countries in the context of the Ukrainian crisis.

In terms of individual economic indicators, all countries of the Baltic region are very different, which confirms the existence of significant differences in the level of industrial and technological development of these states and is a possible basis for their further investment and production cooperation (Mezhevich et al. 2016; Smirnyagin 2011; Baklanov and Shinkovsky 2010). Germany, Sweden and Norway are the most industrialized countries in the Baltic region—they play a role of growth drivers, while the least developed countries are the Baltic states, Poland and Russia. In many ways it is determined by historical aspects (the period of socialist development and the different ways of transition from a planned to a market economy); the time of accession to the European Union (Latvia, Lithuania and Estonia were the last to join the EU) (Komarov 2014; Gusarova 2015; Center for Cross-Border and Interregional Cooperation 2012).

The states of the Baltic region differ in the perception of the borders of Europe and its subregions by European investors (Fedorov and Korneevets 2015; Rietveld 2012). Only to some extent they are due to the nationality of TNCs. For example, Swedish companies often view the Baltic States as a continuation of Northern Europe, while Russian TNCs often make a distinction between the CIS and the EU, attributing Estonia, Latvia and Lithuania to a particular geographic sector quite arbitrarily (Kuznetsov 2015).

According to Siemens, the states of the Baltic region are Finland and the Baltic States; for Daimler, Deutsche Bank, and Henkel, they are Eastern Europe; for BMW and Allianz—Central and Eastern Europe (CEE). The Norwegian bank DNB associates Eastern Europe and, as part of this region, the Baltic countries and Poland with Europe without Norway. The Swedish company Hennes & Mauritz virtually considers each of the Baltic States as separate entities.

For some other companies, the Eastern European subregions form independent geographic sectors. For example, the Swedish Nordea Bank, Telia Sonera and SEB, the Russian oil giant Lukoil and the Finnish company Sampo have a separate sector called "the Baltic countries". Central and Eastern Europe is an independent geographic sector for the Russian Sberbank, the Dutch Heineken and the German RWE.

At the same time, some companies demonstrate quite an atypical segmentation. For example, Gazprom does not attribute the Baltic States to Europe but to the former USSR, which is explained by the specific nature of the gas transportation system inherited from the Soviet period. The Swedish Ericsson considers the Baltic States to be Northern Europe and Central Asia instead of Central or Western Europe. It relates to an enhanced perception of the markets of the nearby countries as the Swedish 'home market'. At the same time, Svenska Handelsbanken classifies the Baltic States as 'other Europe' because the bank began its expansion from Scandinavia to the UK. The Dutch Philips Electronics attributes the Baltic States to

Table 1 The position of the Baltic countries in geographical segments of the 50 leading companies in Sweden, Finland, the Netherlands, Norway, Germany, Russia, and Denmark according to the market capitalisation (as of 31 March 2014)

Company	Capitalisation (billion USD)	Country	Branch of industry	Geographical segment for records or informal region, where the Baltic States are attributed to[a]
Volkswagen	119.2	Germany	Automobile	Europe without Germany
Siemens	118.6	Germany	Electrical and electronic engineering	Europe, CIS, Africa and Middle East (sometimes without Germany)/ Finland and the Baltic States
Bayer	112.1	Germany	Chemical	Others (without Germany, USA and China)/Europe
BASF	102.1	Germany	Chemical	Europe without Germany
Daimler	101.6	Germany	Automobile	Other countries (without Western Europe, Asia and America)/Eastern Europe (including Turkey)
Novo Nordisk	100.8	Denmark	Pharmaceutical	Europe (EU, EFTA and Western Balkans)
SAP	99.6	Germany	IT	Europe, Middle East and Africa (sometimes without Germany and France)
Gazprom	91.3	Russia	Oil and gas	Foreign countries/Former USSR without Russia
Statoil	90.0	Norway	Oil and gas	Eurasia without Norway/Europe without Norway/the world without Norway, Sweden, Denmark and USA
BMW	81.2	Germany	Automobile	Europe without Germany/CEE
Allianz	77.5	Germany	Insurance	Other Europe without the 5 leading countries of western Europe and Switzerland/CEE/growing markets (when indicating linguistic areas in Western Europe)
Deutsche Telecom	72.2	Germany	Mobile communication	Europe without Germany/EU without Germany
Rosneft	70.7	Russia	Oil and gas	Far abroad countries/Europe

Source Kuznetsov 2015

[a]Some TNCs do not have branches in the Baltic countries; if they do not supply goods there, we defined their geographical segment according to the neighbouring countries. Segments 'Europe' and 'Eastern Europe' include Russia unless indicated otherwise

'growing markets.' The German Henkel and other companies have a similar approach, along with others (Table 1).

Let us consider further features of investment and production cooperation between the countries of the Baltic region (Evchenko 2012; Kosov and Gribanova 2016). Investment and production integration in the Baltic region did not start to develop in full until 2004, when Estonia, Lithuania, Latvia and Poland joined the European Union (Komarov 2014). As is known, the integration policy in the EU is carried out in two directions—"downwards" (the EU pan-European programs) and "upwards" (as the programs worked at the regional level). At the same time, the "upwards" integration policy is manifested in various forms (Bolotnikova and Mezhevich 2012; Sologub 2015).

The most important element of the integration policy in industry is foreign direct investment and interaction in the framework of cross-border cooperation. In the process of business internationalization most of the investors seek to develop their business in the neighboring border areas, given the existence of close economic and sociocultural ties (Komarov 2014; Panshin et al. 2014; Fedorov et al. 2012; Rietveld 2012).

The narrow internal market of the post-Soviet Baltic States, characterized by low trade protective barriers, will not stimulate the establishment of production subsidiaries or service departments within a number of TNCs from the neighboring countries. This is clearly illustrated by the example of the 10 main (by the amount

Table 2 The leading Russian non-financial TNCs at year-end 2013

Company	Branch of industry	Foreign assets (million USD)	Significant assets in the Baltic countries
Gazprom	Oil and gas	40,128	37% of each "Lietuvos dujos" and "Amber Grid" in Lithuania; 34% of "Latvijas Gaze" in Latvia; 37% of each "Eesti Gaas" and "Vorguteenus Valdus" in Estonia
VimpelCom	Telecommunications	36,948	–
Lukoil	Oil and gas	32,640	Subsidiaries controlling filling station chains in Lithuania, Latvia and Estonia
Evraz	Steel industry	8715	–
Rosneft	Oil and gas	8399	"Itera Latvija" (subpartner of Gazprom in Latvia and Estonia)
Sovcomflot	Transport	5293	–
Severstal	Steel industry	4784	50.5% of "Severstallat" in Latvia
Rusal	Non-ferrous metal industry	3655	–
RZD (Russian Railways)	Transport	3222	–
Sistema	Conglomerate	2966	–

Source Kuznetsov 2015

of foreign assets) non-financial TNCs from Russia—only 4 of them have subsidiaries in the Baltic States, while only Gazprom and Lukoil set up local distribution companies in Estonia, Latvia and Lithuania (Table 2).

Investment and Production Cooperation Between Russian and Baltic Companies

Let us consider in more detail the specific features of investment and production cooperation between the Baltic States and Russia. For example, Russian capital has strong positions in several sectors of Lithuanian economy. The leading positions are held by Lithuanian companies controlled by Russian companies such as Lukoil-Baltia (a subsidiary of NK Lukoil, which has the largest network of filling stations in the country); the electricity supply company Energijos realizacijos centras (Inter RAO UES); the Kaunas Thermal Power Station and gas importing company Lietuvos Dues (Gazprom); the mineral fertilizer plant Lifosa (EuroChem); a firm for metal structures production Nemunas (Mechel).

Among important Russian investment projects there are such Russian business entities as the fish processing enterprise Vichiunai-Rus, the meat processing plant Kaliningradsky Delikates (Kaliningrad Delicacy), the confectionery shop Nova Ruta, a complex chemical fertilizers plant, and the shopping center Akropolis (Ministry of Foreign Affairs of the Russian Federation 2016).

Russian-Latvian relations are developing dynamically, too. For example, in the Latvian Register of Enterprises, there were 3301 Latvian-Russian joint companies registered. At the same time, in Russia there are about 320 Russian-Latvian joint companies (Russian-Latvian trade and economic relations 2016).

The leading Russian investors in Latvia are Transnefteproduct OJSC—USD 36.55 million (34% of LatRos-Trans LLC shares; operations in the pipeline transport sector and the delivery of steam and hot water); Gazprom OJSC—13.57 million (34% of Latvijas Gaze LLC shares; the main activities are production, processing, and transportation through distribution pipelines and sale of gaseous fuels); Moscow Commercial Bank OJSC—10.82 million (99.87% of Latvijas Biznesa Banka JSC shares; operations in the field of monetary intermediation); private entrepreneur Igor Tsyplakov—10.82 million (100% of Rigensis Bank shares); private entrepreneur Yury Shefler—7.9 million (100% of Meierovica-35 LLC shares; provides accommodation services) (Mezhevich and Sazanovich 2013).

Russian-Estonian economic relations are also developing. For example, among the leading Russian companies traditionally investing in the economy of Estonia, there are Lentransgaz and Gazprom. Among the new companies we can mention Ecomet Invest which in 2014 announced a future launch of a plant for recycling lead acid batteries and the production of lead, polypropylene, sodium sulfate, and ammonium sulfate in Slantsy (Russia).

Emlak, the Russian manufacturer of industrial paints and varnishes, and the manufacturer of organic cosmetics Natura Siberica also invest in Estonia (Nevskaya 2016).

Cross-Border Investment and Production Cooperation Between Russia and the European Union Countries

The development of investment and production cooperation between the companies of Russia and the EU countries continues. For example, the Swedish investments in Russia are made by such large companies as Sandvik, Scania, Volvo, Skanska and by such transnational companies with Swedish capital as ABB, Vostok-nafta, IKEA, Stora Enso and others.

Currently, the Swedish investors in Russia operate in such areas as fuel and energy complex, engineering, telecommunications, construction, trade and production of various consumer goods (Fedorov et al. 2012).

The leading positions in cooperation are held by ABB (18 companies are located in the Russian Federation); AGA (2 plants for industrial gases production); Alfa Laval (a factory for the production of heat exchangers); Assi Doman (cardboard factory); Autoliv (the manufacture of automobile belts and airbags); Fristads (two textile factories); NCC (asphalt plant); PLM (can making); Pripps (co-owner of the Baltika brewery); Sandvik (the owner of Sandvik-Moscow plant of hard alloys); Ericsson (production of telephone exchanges); IKEA (furniture shopping centers, furniture production), MEGA (Export and import, Sweden's trade balance 2017).

The cooperation in software production is developing: the Swedish companies Progate, Texel, Kivicom and ORC Software are working in St. Petersburg. The software of the latter, developed in collaboration with the Russian specialists, is used at the London and Montreal stock exchanges (Baranova 2012; Fedorov et al. 2012; Export and import, Sweden's trade balance 2017).

The main branches of Russian economy in which Finnish investors operate are wood processing (Stora Enso, UPM-Kymmene); food industry (Valio, Atria, Fazer); chemical industry (Nokian Renkaat, Tikkurila, Teknos); the production of engineering structures and insulation materials (Parok); residential, industrial and road construction (YIT, SRV, Lemminkainen); and the energy sector (Fortum). The total number of Finnish companies operating in Russia is about 650.

The main Finnish investors are Fortum Corporation (Nyaganskaya power plant in the Khanty-Mansiysk Autonomous district), Nokian Tyres Corporation (in 2012 commissioned the second car and truck tire factory in Vsevolozhsk) and others (Fedorov et al. 2012). About 2000 companies with Russian capital operate in Finland in consulting and trade-brokering services, tourism, transport and logistics.

For the time being, the largest Russian investments in Finland have been supplemented with Rosatom State Corporation's participation in the construction of the Hanhikivi-1 nuclear power plant. The companies with Russian-Finnish investments include Gazprom, Nizhex Scandinavia (chemical production), OJSC

Nizhnekamskneftekhim, Norilsk Nickel, Arctech Helsinki Shipyard, Cytomed (pharmaceutical products) and others (Embassy of the Russian Federation in Finland 2016).

Positive experience in implementing investment projects involving Polish capital has been gained in several regions of the Russian Federation. For example, the company Pfleiderer Grajewo built a plant for the production of chipboards in the Novgorod region; a timber processing plant was built in the Vologda region (Barlinek company); factory for the production of hygienic products was built in the Moscow region (the investor is the Torun Bandaging Material Plant); a factory for wood depth processing in the Kemerovo region (Wiedemann Polska in cooperation with ZAO Anzher plywood mill) and a number of others (Lisyakevich 2016).

Polish companies are also active in the Russian construction market. In 2011 the company Unibep, with the financial support of a consortium of Polish banks, completed two construction projects in the regions of Russia: the construction of a hotel and an office building "Airport City" in the Leningrad region and the "Vnukovo" hotel in the Moscow region (Mezhevich 2014).

It is planned to create an automotive cluster "INTRALL" with the participation of Russian and Polish companies in the Stavropol Territory, which is to produce up to 60,000 light commercial trucks a year. The Leningrad region and the Lower Silesian voivodeship are planning to launch an investment project for the construction of a "Wholesale Center for Agricultural Products" near St. Petersburg (Russian-Polish Trade and Economic Cooperation 2014).

PESA Bydgoszcz, which produces rolling stock and trams, is another example of a successful company exporting to Russia. A successful Russian investment in Poland is OJSC EuRoPol Gaz, the owner of the Polish section of the Yamal gas pipeline; half of its shares is controlled by Gazprom.

Thus, the corporate integration of Russian and foreign companies in the Baltic region is determined by three factors: (1) the predominance of transit and energy projects; (2) low competitiveness of most branches of Russian industry in comparison with the developed countries of the Baltic region; (3) institutional deficiencies in the dialogue between the EU and the Russian Federation (Evchenko 2012).

Intraregional Features of Economic Integration of the 'Baltic' Countries of the European Union

Investment and production cooperation between the countries of the European Union is actively developing in the Baltic region. For example, the number of Polish companies operating in Germany is about 20,000 of which 4000 are located in Berlin. About 100 branches of Polish construction companies (Budimex, Kopex, Polservice, and others) operate in the territory of Germany.

The influential Polish investors working in the German markets include such companies as Orlen, Boryszew S.A., ComArch, Selena, Unimil Sanplast S.A., Smyk, PGNiG (German Mining Company of Oil and Gas) (Kryzhanovskaya 2008; Kuznetsov 2015).

The Finnish-Estonian investment and production relations in the Baltic region are expanding. For example, in Estonia there are approximately 4500 Finnish companies, and in Finland there are about 3500 Estonian companies (DELFI 2013). There are 51 Estonian subsidiaries operating in Finland. The most famous of them are Tallink AS (Tallink Silja OY), AS Tavid (Tavex OY), AS Harju Elekter (Satmatic OY), Uptime, Bigbank, LHV, BLRT, Eesti Energia (Solidus OY), Rand & Tuulberg and others.

Finnish corporation VALIO has 15 factories in Finland, two plants in Estonia and a production facility in Belgium that produces packaging materials. The Fazer Corporation which produces confectionery and bakery products has more than 1400 restaurants and cafes in Finland, Norway, Denmark, Estonia, Sweden, Latvia and Russia.

Thus, analyzing the data of UNCTAD, the major investors in the territory of the Baltic region are Germany, Finland, Sweden, Denmark and Poland. The reasons for the leading position of these countries are obvious—developed industry, the presence of large TNCs, strong ties with other EU countries. The presence of Poland in this list is determined by the progressive growth of its economy and large investment projects in the territory of the state.

The so-called "neighborhood effect" is mostly manifested in the cross-border investment cooperation within the triangle of Lithuania-Estonia-Latvia, whose joint development was largely determined by specialization within the planned economy of the USSR. Another factor for economic and industrial integration is the technological specificity of industrial production in these countries (Akhutina et al. 2013; Panshin et al. 2014).

In general, the policy of cooperation and investment is mainly concentrated in the Nordic countries—Sweden, Denmark and Finland. Moreover, these processes are determined by the desire to create a large interstate market. In the present context of economic integration and globalization, the private capital of the Northern Baltic is mostly controlled by Swedish investors. The policy of Sweden and Finland to expand the market for its products led to joint investment in the Baltic countries—Lithuania, Latvia and Estonia.

It should be noted that the list of integration and cooperation mechanisms in the Baltic region is not complete (Akhutina et al. 2013; Baranova 2012; Gumenyuk, Kuznetsova and Osmolovskaya 2016; Evchenko 2012). Another successful mechanism for cooperation is the establishment of control over regional TNCs through the foundation of subsidiaries and their sale.

The economic crisis of 2008–2009 affected the economic policies of the countries in the Baltic region—unemployment increased, and investment attractiveness of the Baltic countries decreased. At the same time, the largest TNCs in Germany, Sweden, Denmark and Finland tend to generate innovations in the main enterprises in their countries and establish their subsidiaries in Asian countries where the labor is cheaper than in the Baltic States and the market for industrial products is growing.

The joint projects between Germany and the Scandinavian countries are good examples of successful long-term cooperation. The flow of mutual investments between Germany, Denmark, and Sweden has been continuing since the 1970s. This contributed to the industrial and territorial development of the border areas of these countries (Panshin et al. 2014; Sologub 2015).

The most successful example of the policy of industrial and economic integration among the countries of the Eastern Baltic is Poland. A large and receptive domestic market, the availability of relatively cheap labor compared to the Scandinavian countries and stable legislation attracted big investors and created cross-border cooperation zones on its territory (Panshin et al. 2014). Poland itself is expanding its market due to the cooperation with the post-Soviet countries, in particular with Lithuania.

Conclusion

The study showed the differences in perception of European borders by the business communities. Only partly they are explained by the nationality of TNCs. The different perception of parts of Europe by individual TNCs, including representatives of different sectors, seems to be reflected in the positions of European businessmen in resolving problems in relations between Russia and the EU, which will require further investigation as the "war of sanctions" develops.

Russia's cooperation with foreign countries in the Baltic Sea region has a significant potential in providing a balance in the energy trade, production cooperation and modernization of Russian economy.

Foreign countries of the Baltic region have already established multilateral formats of cross-border cooperation, which significantly accelerated the process of regionalization. The Baltic region has become an important part of global integration processes. The processes of localization are reflected in cross-border cooperation projects and the programs for sustainable development of the Baltic Sea area. New spatial forms of international investment and production cooperation are developing around the Baltic Sea.

Cross-border and trans-border cooperation is of great importance for Russia and her neighbors. It concerns the development of national TNCs and SMEs, the improvement of transport, logistics and customs infrastructure, the introduction of best practices in the implementation of projects in the North-West Federal District of Russia, the development of industrial cooperation and the support of cluster initiatives.

In this context, the main task for all participants should be the maintenance of the achieved level of cross-border cooperation. Interaction with the Baltic countries in the format of Euroregions as well as implementation of bilateral strategies of cross-border cooperation objectively remain a resource for maintaining working relations in the areas of mutual interest and an important tool for smoothing different problems in relations between the countries of the Baltic region. Therefore, the experience of transnational and cross-border cooperation in the Baltic region should be actively studied and should possibly be applied in other regions of the world.

References

Akhutina D, Vorontsova S, Lazovsky S et al (2013) Baltiiskoe more: ot koordinatsii strategii k protsvetaniyu makroregiona (The Baltic sea: from the coordination of strategies to the prosperity of the macroregion). St. Petersburg (in Russ.)

Baklanov P, Shinkovsky M (2010) Transgranichnyi region: ponyatie, sushchnost', forma (Transboundary region: concept, essence, form). Dal'nauka, Vladivostok (in Russ.)

Baranova Y (2012) O mezhdunarodnom sotrudnichestve Severo-zapada Rossii v innovatsionnoi sfere na Baltike (On international cooperation of the North-West of Russia in the innovation sphere in the Baltic). Baltic region. https://doi.org/10.5922/2079-8555-2012-4-12 (in Russ.)

Bolotnikova E, Mezhevich N (2012) "Severnoe izmerenie" i strategiya evropeiskogo soyuza dlya Regiona Baltiiskogo morya ("Northern Dimension" and the European Union Strategy for the Baltic sea region). Pskov Reg J 13:37–47 (in Russ.)

Center for Cross-Border and Interregional Cooperation of the Higher School of Economics (2012) Promezhutochnyi otchet o provedenii issledovaniya «Region Baltiiskogo morya v fokuse strategii razvitiya Evropeiskogo soyuza i Rossiiskoi Federatsii» (Interim report on the study "The Baltic sea region in focus of EU and RF development strategies"). http://www.n-west.ru/wp-content/uploads/2000/02/Promezhutochnyiy-otchyot-issledovaniya_21.03.2012.pdf. Accessed 10 Aug 2017 (in Russ.)

DELFI (2013) The premieres of Estonia and Finland discussed their countries' cooperation in the field of electronic technologies. http://rus.delfi.ee/daily/estonia. Accessed 30 Aug 2017

Embassy of Russia in Poland (2014) Rossiya-Pol'sha. Torgovo-ekonomicheskoe sotrudnichestvo (Russian-Polish trade and economic cooperation). https://poland.mid.ru/torgovo-ekonomiceskoe-sotrudnicestvo. Accessed 30 Aug 2017 (in Russ.)

Embassy of the Russian Federation in Finland (2016). http://helsinki.mid.ru/glavnaa. Accessed 30 Aug 2017 (in Russ.)

Embassy of the Russian Federation in the Republic of Latvia (2016) Rossiisko-Latviiskie torgovo-ekonomicheskie otnosheniya (Russian-Latvian trade and economic relations). https://latvia.mid.ru/rossijsko-latvijskie-torgovo-ekonomiceskie-otnosenia. Accessed 30 Aug 2017 (in Russ.)

Evchenko N (2012) Prigranichnoe sotrudnichestvo Rossii: zakonodatel'naya baza i napravleniya razvitiya (Cross-border cooperation of Russia: legislative base and directions of its improvement). Reg Econ S Russ 13:18–25 (in Russ.)

Export and import, Sweden's trade balance (2017). https://svspb.netsvspb.net/sverige/export-import-saldo.php. Accessed 10 Aug 2017

Fedorov G, Korneevets V (2015) Socio-economic typology of the coastal regions of Russia. Baltic region. https://doi.org/10.5922/2079-8555-2015-4-7

Fedorov G, Voloshenko E, Mikhailova A, Osmolovskaya L, Fedorov D (2012) Territorial differences of innovative development of Sweden. Baltic region, Finland and the North-West Federal District of the Russian Federation. https://doi.org/10.5922/2079-8555-2012-3-6

Gumenyuk I, Kuznetsova T, Osmolovskaya L (2016) Local border traffic as an efficient tool for developing cross-border cooperation. Baltic region. https://doi.org/10.5922/2079-8555-2016-1-6

Gusarova V (2015) Analiz urovnya ekonomicheskogo razvitiya gosudarstv Regiona Baltiiskogo moray (An analysis of the level of economic development of the states of the Baltic sea region). Vestnik Pskovskogo gosudarstvennogo universiteta. Seriya: Ekonomika. Pravo. Upravlenie. https://cyberleninka.ru/article/n/analiz-urovnya-ekonomicheskogo-razvitiya-gosudarstv-regiona-baltiyskogo-morya. Accessed 10 Aug 2017 (in Russ.)

Komarov A (2014) Baltiiskoe sosedstvo: Rossiya, Shvetsiya, strany Baltii na fone epokh i sobytii XIX–XXI vv. (The Baltic neighborhood. Russia, Sweden, the Baltic States against the background of the epochs and events of the XIX-XXI centuries). LENAND, Moscow (in Russ.)

Kosov Y, Gribanova G (2016) EU strategy for the Baltic sea region: challenges and perspectives of international cooperation. Balt Reg. https://doi.org/10.5922/2079-8555-2016-2-3

Kryzhanovskaya V (2008) Ekonomicheskoe sotrudnichestvo mezhdu Germaniei i Pol'shei (Economic cooperation between Germany and Poland). http://www.rusnauka.com. Accessed 30 Aug 2017 (in Russ.)

Kuznetsov A (2015) The place of the Baltic countries in the geographical segments of the largest TNCs in Europe. https://journals.kantiana.ru. Accessed 24 Aug 2017

Lisyakevich R (2016) Geoekonomika v otnosheniyakh Pol'shi i Rossii (Geoeconomics in the relations of Poland and Russia). http://visegradeurope.ru. Accessed 30 Aug 2017 (in Russ.)

Mezhevich N (2014) Vneshnyaya politika gosudarstv Pribaltiki i krupnye infrastrukturnye proekty 2010–2014 godov (Foreign policy of the Baltic states and major infrastructure projects of 2010-2014). https://cyberleninka.ru/article/n/vneshnyaya-politika-gosudarstv-pribaltiki-i-krupnye-infrastrukturnye-proekty-2010–2014-godov. Accessed 10 Aug 2017 (in Russ.)

Mezhevich N, Kretinin G, Fedorov G (2016) Economic and geographical structure of the Baltic sea region. Balt Reg https://doi.org/10.5922/2079-8555-2016-3-1

Mezhevich N, Sazanovich L (2013) Sovremennye problemy rossiisko-latviiskikh otnoshenii (Modern problems of Russian-Latvian relations). Baltic Reg 3(17):93–106 (in Russ.)

Ministry of Foreign Affairs of the Russian Federation (2016) Russian-Lithuanian relations. http://www.mid.ru. Accessed 30 Aug 2017 (in Russ.)

Nevskaya A (2016) Foreign economic relations between Russia and Estonia in modern conditions. http://ecfor.ru. Accessed 30 Aug 2017 (in Russ.)

News from Poland (2014) Russian capital in Poland is still frightening. But not always. http://www.polska-kaliningrad.ru. Accessed 30 Aug 2017 (in Russ.)

Panshin I, Yares O, Zemskova M (2014) Integratsionnoe razvitie prigranichnykh rossiiskikh regionov v ramkakh tamozhennogo soyuza: resursoobespechennost' i ekonomicheskii rost' (Integration development of border Russian regions within the framework of the customs union: resource availability and economic growth). Actual Probl Econ 8(158):204–214 (in Russ.)

Rietveld P (2012) Barrier effects of borders: implications for border-crossing infrastructures. EJTIR 12(2):150–166

Smirnyagin L (2011) Megaregiony kak novaya forma territorial'noi organizatsii obshchestva (Megregions as a new form of territorial organization of society). Vestnik Mosk. University. Ser. 5. Geogr 1:9–15 (in Russ.)

Sologub A (2015) Osobennosti podkhoda Evropeiskogo soyuza k podderzhke mezhdunarodnykh proektov v Regione Baltiiskogo morya (Features of the European Union's approach to supporting international projects in the Baltic sea region). Issues Theory Pract Tambov Diploma 7(57):161–164 (in Russ.)

Everyday Life in the Russian Borderland

Maria V. Zotova, Anton A. Gritsenko and Alexander B. Sebentsov

Abstract

In connecting people's expectances and reality, border crossing has certainly a considerable impact on human behavior and mind, especially in case if this act becomes an individual's regular practice and/or a mean of existence. In this case, borderland can be transformed into a space of everyday interactions between the people and radically changes their life. The authors try to summarize the definitions of everyday life and theoretical approaches to its study used by different experts. The hypothesis is that everyday practice does not simply show distinctions or similarities of the people living close to the boundary on its both sides but allows understanding the shaping of different socio-cultural spaces and the differentiation of border regions as a whole. The authors estimate the opportunities and constraints created by the boundary and motivating its crossing, analyze its reasons and directions at all sections of Russian land borders. They focus on the socio-economic and socio-cultural discrepancies, and the features of neighbouring areas stimulating specific practices distinguishing borderlands from the rest of the state territory.

M. V. Zotova (✉) · A. A. Gritsenko · A. B. Sebentsov
Center for Geopolitical Studies, Institute of Geography, Russian Academy of Science, Moscow, Russia
e-mail: zotova@igras.ru

A. A. Gritsenko
e-mail: antgritsenko@igras.ru

A. B. Sebentsov
e-mail: asebentsov@igras.ru

A. A. Gritsenko
Immanuel Kant Baltic Federal University, Kaliningrad, Russia

© Springer Nature Switzerland AG 2020
G. Fedorov et al. (eds.), *Baltic Region—The Region of Cooperation*,
Springer Proceedings in Earth and Environmental Sciences,
https://doi.org/10.1007/978-3-030-14519-4_9

73

Keywords
The state border · Everyday life · Social practices · Russian borderlands

Borderland belong to those few places where governmental policy and a low ranker's life are closely intertwined, and cross-border practices become resulting output of the activities of social systems at different levels—from the state and society as a whole to near-border or cross-border communities. When socioeconomic order is stable and firmly established, the life of borderland communities complying with worked-out formal and informal rules is equally well-ordered. But what happens when routine order collapses? How local people adapt to arising changes which re-determine the character of interaction with neighbors across the border?

Foreign-policy developments and (geo)political shifts in the relations between Russia and neighboring states had an immediate influence upon people's cross-border practices changing their intensity, character, directions, and re-determining for Russia's borderlands inhabitants the balance between benefit and cost, border-crossing motives and impediments.

The ambiguity of processes going on at the border, when rivalries force to cooperate and the cooperation is performed "over the barriers" (Vendina and Kolosov 2007), makes it necessary to focus on everyday life, to study details, causes of border crossing, labor migrations, shopping and leisure practices, etc. We interpret cross-border practices as human activities relating to border crossing in a broad sense including habits of people, their behavior and agendas. A task of the paper is to describe as far as possible the diversity of social practices emerging near the state border, to find purposes, motives, and factors of cross-border movements in different places, and to determine their significance for the life in Russia's border areas. The paper is based on the materials of numerous field studies conducted in 2010–2017 in border regions of Russia and neighbouring countries resulted in more than 600 interviews with experts.

Living Conditions Near the Border and Cross-Border Practices

The state border is intrinsically dialectical. On the one hand, the border guarantees security; on the other hand, it attracts criminality. Functioning as a filter, the border should theoretically prevent distribution of "negative" practices (illegal migrations, contraband, criminal activities, etc.), in reality, however, it rather encourages them. Besides that, the border is simultaneously characterized both as a barrier and a contact zone. Any border is a limit or barrier dividing territories. The presence of a border inherently complicates interactions, transportation of people and goods, economic relations because its crossing costs much time, energy, and financial resources. The interest of local communities in maintenance of good neighborhood, cross-border links to friends, relatives, and business partners obviously conflicts with many border functions ranging from provision of state sovereignty and security to formation/maintenance of national identity.

At the same time, the border due to its properties not only entails restrictions but also creates some opportunities and everyday practices which use advantages of neighborhood and are not characteristic of the country's inner regions. Everyday life of people who live near the border undergoes transformations under its influence. The social life in border areas develops to a large extent thanks to the existence and nearness of "neighbors" (Brednikova 2008). The borderland is a changeable, flexible space where governmental and private issues, "us" and "them" overlap each other, where social processes go on in different directions and ambiguously.

Reasoning on the influence of border upon everyday life not always is possible in distinct terms of restrictions and opportunities, because just their fusion creates specific conditions of cross-border activities and determines problems to be resolved. A demonstrative example is the realization of common projects in border areas with the participation of both sides. Their proponents and active participants are doubtless beneficiaries of border openness which brings them both financial and moral dividends. However, there would be no need of such activities, if there were no problems and restrictions generated by borders.

Farther, let us consider what opportunities and restrictions the border creates as well as what cross-border practices emerge under the influence of its properties (barrier and contact functions, regime of functioning) and of the properties of neighboring countries' adjacent regions and mutual relations between them (differences in prices, business rules and conditions, population distribution patterns, etc.). By the example of various situations on Russia's borders, three main spheres (those that unify practices resulting from similar factors and the same motives) are considered in details: (1) market of services; (2) labor and unreported employment; (3) commerce and leisure.

Market of services. One of characteristic features of borderland is its peripheralness and remoteness from developed markets of services. The topology and transport connectivity of territories have frequently such parameters that for many border areas the nearest economical and cultural centers are situated abroad. It is more convenient and quicker for inhabitants of some border districts of Smolensk *oblast'* to come to Vitebsk or Mogilev than to Smolensk as well as for those of Bryansk *oblast'* to Gomel'. In Kaliningrad *oblast'* such an alternative is provided by capitals of nearest voivodeships in Poland while in Pskov *oblast'* by capitals of municipalities in Latvia. It is more convenient for border areas inhabitants of West Kazakhstan Region to travel to Orenburg or Samara while for those of Atyrau Region to Astrakhan. This is obviously inappropriate but could be less time-consuming and less expensive that to resolve emerged problems in one's own country. Border crossing becomes a method to compensate costs. For example, international airports of Gdańsk, Tallinn, Riga, Vitebsk, Heihe, Harbin, Ulaanbaatar accommodating low coaster flights, are used by many people from neighboring border regions.

High-level medical services, especially in the spheres of cardiology, surgery, endoprosthesis replacements, and dentistry, it is difficult to receive in middle and small border towns. Their inhabitants are in a dilemma where to go: either to one's "own" medical facility or to a foreign one. The choice depends upon many private reasons including inherent dangers and fears, but it is a fact, however, that people of Kaliningrad, Smolensk, and Pskov *oblasts* seek medical services in near-border centers of Belarus, Estonia, Latvia, and Poland increasingly frequently. Even such a term as "birth tourism" has appeared here. Another variety of medical tourism—dental one—develops near the border with China where dentist's services are less expensive and of higher quality (Ryzhova 2013). At the border between Russia and Kazakhstan, it is common to go to Russia for medical services (births, open heart surgeries, diagnosis and treatment of oncological diseases). According to researchers of Orenburg State University, only the Vision Correction Center in Orenburg receives for treatment about 4 thousand Kazakhstan citizens annually.

Russia's citizens realize actively the opportunities to study abroad at neighboring European universities, and recently also at Chinese ones. Higher education institutions beyond the border attract by their closeness to home and moderate tuition fees. Offering "good education" from the point of view of local people, they are quite able to meet competition with Russian universities situated in large scientific and educational centers or capitals. Their advantages are: (a) closeness to home (possibility to see parents frequently); (b) education and accommodation costs affordable for the parents of a student; and (c) less severe admission requirements. According to many local experts, such a choice is a lot of "weak" school-leaving certificate holders and children from low-income families.

There are also other purposes. For example, school leavers from Karelia enter the universities of Kuopio, Mikkeli, Lappeenranta, Imatra, and Joensuu increasingly frequently while people from Kaliningrad choose higher educational institutions of Gdańsk and Sopot and not those of their own city. According to our interviews with students in Kaliningrad, their striving to move to another country are not determined by actual difficulties and poor perspectives at home, but is a result of expectations to find later a well-paid and interesting job in the profession there, where, as they think, this is "guaranteed for all". University studies in the EU countries are regarded as a first step in this way. Availability of a diploma recognized in Western states seems to be a starting condition for a fruitful career and life as a whole.

The situation in Russia-Kazakhstan and Russia-Belarus border areas is opposite. People from Kazakhstan and Belarus traditionally come to study to neighboring Russian cities (Omsk, Orenburg, Smolensk). According to local experts, the number of Belarus citizens, who study in three Smolensk's leading higher education institutions, exceeds 1500 (70% of them came from Vitebsk *oblast'*, and 20% from Mogilev *oblast'*) while there are over thousand Kazakhstan students in Orenburg and Omsk state universities. Education of children in large centers of the neighboring country becomes one of multipurpose long-term life strategies of population.

Labor market and illegal employment. Liberalization of border regime at some Russian boundaries, especially the introduction of practice of local border traffic (LBT) and simplification of visa-issuance procedure, has provided for border areas residents new earning opportunities ranging from *legal employment* at large plants abroad to smuggling as a form of self-employment.

At the border between Finland and Karelia, due to differences in resources availability and conditions of economic activities, a trans-border production complex has emerged which includes logging at the Russian side and subsequent timber conversion in Finland. Despite the fact that such cooperation was asymmetric in its character, it created jobs in peripheral areas of the Republic of Karelia and exerted a profound influence upon its economy and social life.

In the Far East the Chinese labor force is widely used in the economy, first of all in construction and agriculture. Chinese businessmen establish their firms in the Russian territory (Mishchuk 2016). The Chinese business is characterized by a high degree of isolation and closeness, it creates labor market mainly for its own people. Russian residents have practically no access to such enterprises.

After the USSR disintegration, labor links between closely situated border towns and regions began to be weaker rather quickly. At the borders with the Baltic states, the strengthening of boundary regimes, long waiting at border crossing points as well as difficulties with legal employment of foreigners resulted in the situation when permanents jobs were replaced by unreported and seasonal employment. Some cases of cross-border employment remained at Russian- Ukraine boundary up to the recent time. Residents of Ukraine worked on a massive scale at sugar plants in Kursk *oblast'* as well as in coalmines in Rostov *oblast'*. Up to the early 2000s, workers from towns of neighboring Lugansk *oblast'* commuted daily to the coalmines of Gukovo where salaries were higher than in Ukraine. "Delivery of foreign labor force" was minimized because of administrative difficulties (Kolosov 2016).

After the events of 2014, one-way crossing dominates at the border between Russia and Ukraine: about 80% of those who cross the boundary are Ukrainian citizens coming to Russia to earn money.[1] The bulk of labor migrants settles in Belgorod *oblast'* within the areas most developed economically. Ukrainians as a rule, prefer temporary jobs, because they plan to return to Ukraine while Russian citizenship and official registration would make travels to homeland significantly more difficult.

Informal kinds of labor activities, such as shuttle trade and shadow economy, are also widespread at the border. They include deliveries of cheap alcohol and sweets from Kazakhstan, seafood—in the Far East, vodka and cigarettes—from Kaliningrad to Poland and from Donetsk People's Republic and Lugansk People's Republic to the Russian Federation, as well as unofficial procedures of timber export to China.

---

[1]According to the Head of the Department of International Relations and Border Cooperation of Belgorod *oblast'*, January 2016.

Despite a notable harm to the economy of border areas and damage inflicted to some local enterprises especially in the field of food industry, shadow business is the source of income for a significant part of local population. This is not only a personal "safety bag", but also a factor providing socioeconomic stability in borderlands, self-employment of population, and maintaining of habitual living standards. All this forces local administrations to turn a blind eye on those things, which are regarded as unwarrantable by regional or federal authorities. This motivation concerns also numerous intermediary services for completion of various paperwork including international certificates of insurance necessary for visits to neighboring country; consulting and legal services for businessmen who start up; interpreting and translation services (especially at the Russian-Chinese border); services of realtors for purchasing real estates in neighboring country (mainly in Finland and Estonia).

The border diversifies and modifies labor market creating jobs for local residents (border guard, customs) and attracting people from abroad, from border areas of neighboring states. While generating cross-border labor market, the border stimulates also local labor market which results in the development of the objects highly demanded under specific conditions. Thus, it is very notable that at the border with the EU countries there are significantly more pharmacies and filling stations than in other borderlands.

Consumption and leisure. The life near the border changes the consumption character of local people notably. Difference in prices permits to cut down expenses, and rich assortment increases satisfaction with life. Consumptive tourism became one of the most popular motives for border crossing, especially at Russia's boundaries with the EU countries and China (Ryzhova 2013). According to the opinion of local residents, goods bought in Europe possess higher consumptive qualities that their Russian equivalents. People in Kaliningrad *oblast'* believe that sausages sold in a Polish discounter contain more meat than those in a local shop. Unlike Russian citizens who think that European products are better than domestic ones, the Finns and the Poles, however, do not believe in the quality of Russian food (milk, meat) and the appropriate level of sanitary control, that is why even the difference in prices does not contribute to the increase of consumptive demand from their side. At the same time the Kazakhs and the Chinese underline the high quality of some Russian products (beer, chocolate, sweets) while comparing them with their local equivalents. The Chinese think that Russian food contains less pesticides. Besides that, many Russian goods (flour, dairy produce, condensed milk, yogurts) are highly demanded by the Chinese who regard them as exotic.

Directions of cross-border shopping traffic depend very much on economic situation and exchange rate fluctuations. Thus, before 2014 when not only the Ukrainian crisis has broken out but also oil prices dropped entailing subsequent fall of the rouble, there were mainly Russian citizens who made shopping tours abroad, but the end of 2014 was characterized by a sharp increase of traffic to Russia from many neighboring countries, first of all China, Kazakhstan, Belarus, Azerbaijan, and the Baltic states. In Blagoveshchensk the number of Chinese shopping tourists went up two-and-a-half times during the first half of 2015.[2]

At the same time the flow of Russian borderlands residents to the neighboring regions of the EU and China has decreased notably but not cardinally. Well-established habit for consumption abroad took over. Even the abolishment of LBT regime by the Polish side in July 2016, resulting in additional difficulties with visas for the Russians and a fall in consumer demand in near-border supermarkets for the Poles, did not lead to a cardinal change in the situation. The expenses of Kaliningrad residents for shopping on the other side of the border dropped as little as by 15.3% in the third quarter of 2016 in comparison with the same period of 2015 (Razowski 2017). At the same time, the number of the Poles coming to Kaliningrad *oblast'* to buy fuel has declined sharply. It is a clear example that strong motives stimulate to overcome obstacles while not very strong ones inspire to search for alternatives.

Economic motives for combination of shopping and relaxation can principally differ on both sides of the border. Thus, feeling the need to change their environment, the Russians go to Finland or Poland mainly to spend money while the Finns or the Poles go to Russia rather to save it taking into account significant difference in services costs (Izotov and Laine 2012). For the Russians economical reasons become more essential at the borders between Russia and Kazakhstan as well as between Russia and China (Simutina and Ryzhova 2007). These differences in behavior as well as the very image of tipsy people wasting money for bagatelles, on the one hand, and of misers who count pennies, on the other hand, generate negative ethnic stereotypes among *experts* in the field of contacts with neighboring countries and later among all others.

Conclusion

The border transformation from a line (stripe) dividing states into the space of intensive contacts and interactions of people changes their life cardinally. Residents of borderlands on both sides of the boundary begin to make profit of the neighborhood, namely of the difference in prices, opportunities, and life standards. Even those, who in no way participate in cross-border movements, exchanges, and communications, receive their small part of border-related profits which are redistributed from commuters to their closest circle members as well as to other spheres of local economy.

---

[2]Rambler finance (2008) Glava Primor'ya: chislo kitayskikh turistov v kraye uvelichilos' v 2,5 raza (The head of Primorye: the number of Chinese tourists in the province increased by 2.5 times). https://finance.rambler.ru/news/1676585/https://finance.rambler.ru/news/1676585/. Accessed 10 November 2017.

Lack of incentives to cross the boundary can be significantly more important than obstacles to overcome while crossing state borders. When motivation is weak, the abolition of barriers does not change anything, whereas strong incentives force to include the border factor into the set of inevitable social costs.

Spontaneous speculative transactions ("to buy cheap—to sell at a high price") "sewed" together transected areas and maintained the achieved living standards of local people. This kind of transactions typical of the first post-Soviet decade was later replaced by more diverse interactions related to the development of small business, services, and tourism. Nevertheless, the difference in prices continues to be an important incentive for cross-border interchange at both "old" and "new" borders of Russia. The changes in practices based on family, friendship, and professional ties were equally strong. The generations, bound by the common Soviet past, studies in universities, work at industrial facilities, lose their consolidated positions and quit the scene as they become older. Their successors are driven by other motives and interests, they are oriented at other emotional and pragmatic landmarks. This type of cross-border interactions, however, continues to be important as before thanks to intermarriages and new waves of migrations to neighboring countries. The social practices, which compensate the deficiency of quality services and create a kind of symbiotic relationships between borderlands communities, remain most stable.

Despite all the diversity of situations at Russia's boundaries, the cross-border practices form a common background uniting all borderlands. On the one hand, these practices are *very flexible* and can change quickly in dependence on international situation and bilateral relations, the state of economy in adjacent countries and regions, exchange rates, prices, customs duties, transport tariffs, etc.; on the other hand they are *very stable*. They are so well-established in the life of boundary communities that people can not live anymore without realizing advantages of their near-border position; even significant changes in border regime does not exert a notable influence on them.

Crisis developments and acute international conflicts do not entail a total stop of cross-border interactions. Daily living needs of people, requirements of economy, and positive cross-border cooperation experience accumulated over the previous years facilitate maintaining of critically needed level of contacts. The attitude to a "good neighbor" is subject to situational changes to a lesser extent than the attitude to a neighboring state which alternates from plus to minus easily.

Partnership between countries creates certain prerequisites for activation of local cooperation, but does not always lead to the growth of its intensity. A characteristic example is the border between Russia and Kazakhstan where the processes of boundary "banalization" are significantly less prominent than in Russian-Polish or Russian-Chinese borderlands. The causes are: low density of economic activities, long distances between large cities, poor infrastructure, and low mobility of population.

Acknowledgements "This work was done under the state assignment of the Institute of Geography RAS (0148-2019-0008, AAAA-A19-119022190170-1)". The article was prepared with partial support of the Immanuel Kant Baltic Federal University, the Russian Academic Excellence Project (5-100).

Ethical Approval and Informed Consent All the procedures performed in the study conforms to ethical standards applicable to conducting this kind of research including fieldwork collection of sociological information (Commission Decision of the Institute of Geography, Russian Academy of Sciences on conformity of an empirical research project to ethical standards, No. 3 from April 1, 2014). Informed consent was verbally obtained from all individual participants (experts and informants) included in the study.

References

Brednikova OE (2008) Prigranich'e kak social'nyj fenomen (napravlenija sociologicheskoj konceptualizacii) (Borderland as a social phenomenon (directions of sociological conceptualization). Vestn Sankt-Peterbg Univ 12(4):492–497 (in Russ.)

Izotov A, Laine J (2012) Constructing (Un)familiarity: role of tourism in identity and region building at the Finnish-Russian border. Eur Plan Stud 21(1):93–111

Kolosov VA (2016) Transgranichnaja regionalizacija i frontal'erskie migracii: evropejskij opyt dlja Rossii? (Cross-border regionalization and frontal migration: european experience for Russia?). Reg issled 3(53):83–93 (in Russ.)

Mishchuk SN (2016) Rossijsko-kitajskoe sotrudnichestvo v sel'skom hozjajstve Dal'nego Vostoka Rossii (Russian-Chinese cooperation in the agriculture of the Russian Far East). Izvest RAN Ser Geogr 1:38–48 (in Russ.)

Razowski Ł (2017) Pół roku bez małego ruchu granicznego. Resort podsumowuje zyski i straty. In: Olsztyn Onet. http://olsztyn.onet.pl/pol-roku-bez-malego-ruchu-granicznego-resort-podsumowuje-zyski-i-straty/196zwdm. Accessed 20 Mar 2017

Ryzhova NP (2013) Jekonomicheskaja integracija prigranichnyh regionov (Economic integration of border regions). IJEI DVO RAN, Habarovsk (in Russ.)

Simutina NL, Ryzhova NP (2007) Jekonomicheskie i social'nye vzaimodejstvija na transgranich-nom prostranstve Blagoveshhensk-Hjejhje (Economic and social interactions on the transborder area of Blagoveshchensk-Heihe). Vestn DVO RAN 5:130–144 (in Russ.)

Vendina OI, Kolosov VA (2007) Partnerstvo v obhod bar'erov (Partnership to bypass barriers). Rossija v global'noj politike 5(1):142–154 (in Russ.)

The Role of Tourism in Cross-Border Region Formation in the Baltic Region

Elena G. Kropinova⬤

Abstract

Today, with the rise in confrontation between Russian and the EU, the role of tourism as a peaceful form of cooperation is of special interest. People-to-people cooperation emerging in the process of service provision could shift the interactions towards a more positive way. This paper focuses on tourism flows between regions of different countries in the Baltic region. Sometimes such flows result in the region formation. The aim of this study is to examine the trend of cross-border region formation in the Baltic and to highlight the role of tourism in this process. Data and information were collected during the author's visits to the Baltic region countries, and a study of Lithuania-Russia and Poland-Russia cross-border programme documents was conducted for the critical analysis of the situation. The results show that the regional formation is at various levels of development in different parts of the Baltic region. Following an in-depth multi-dimensional analysis of research results, the recommendations for national and regional authorities for strengthening the interregional cooperation were developed. They may be useful in developing strategies for the development of border regions (often remote from the center municipalities) as well as tourism and recreation, considering possible interactions with the subjects of neighboring countries for the purposeful establishment of cross-border tourist and recreational regions.

Keywords

Cross-border regions · Tourism · Regional development · The Kaliningrad region · The Baltic region

E. G. Kropinova (✉)
Immanuel Kant Baltic Federal University, Kaliningrad, Russia
e-mail: kropinova@mail.ru

G. Fedorov et al. (eds.), *Baltic Region—The Region of Cooperation*,
Springer Proceedings in Earth and Environmental Sciences,
https://doi.org/10.1007/978-3-030-14519-4_10

Introduction

Cross-border regional formation, actively developing in the era of globalization, captures various types of economic activity. Tourism and recreation, which are developing rapidly in the last half-century, are becoming more actively involved in this process.

Tourist and recreational components are available in many "complex" cross-border regions, which arise due to the interaction of economic entities of various economic activities at different territorial levels. Along with the "complex", the "sectoral" and the "cross-sectoral" cross-border regions are formed, which arise in the interaction of economic entities of one or two - three types of economic activities. These include cross-border tourist and recreational regions, especially actively forming within the European Union, where the contact function of the border is particularly great while the barrier function is insignificant. Russia only borders with the EU countries in the Baltic region, and here the processes of cross-border region formation with the participation of Russian regions are quite intensive, including in the sphere of tourism and recreation. Studies conducted at the Immanuel Kant Baltic Federal University, including those under the leadership and/or direct participation of the author, made it possible to make various practical, theoretical and methodological conclusions about the processes of formation of cross-border tourist-recreational regions in the Baltic macroregion as a whole, as well as inside it at meso- and micro-level.

Approaches to Studying Cross-Border Regions

Cross-border research in the field of tourism and recreation began in the EU and North America with the growth of transparency of state borders as one of the components of research in the formation of complex transboundary regions. Particular attention to tourism and recreation in studying institutionalized cross-border regions which are formed in the EU and on its borders (Euroregions) is given by Brim (2013), Grama (2011), Kovacs (2011), Gal (2009), Scott and Laine (2012), Studzieniecki et al. (2016), etc. The works by Lagiewski and Revelas (2004), Vodeb (2010) are devoted to cross-border cooperation in tourism, Wachowiak (2006), Prokkola (2008), Timothy (1999) specialize on cross-border cooperation in tourism; and cross-border tourist destinations are studied by Seric and Marcovic (2011), and others.

In Russia, the formation of complex transboundary regions has been studied since the mid-1990s. These studies are based on a systematic approach to the investigation of spatial localization of interacting inter-entities in the border areas of neighboring countries. A great contribution to this line of research was made by Baklanov (2014), Vardomskiy (2017), Baklanov and Hansey (2004), Gerasimenko and Gladkiy (2005), Gerasimenko (2012), Katrovsky (2015), Kolosov and Vendina (2011), Kolosov and Turovsky (1999), Korneevets (2010a, 2010b), Manakov (2014), Mezhevich, Kretinin and Fedorov (2016), Turovsky (2004), Fedorov (2016).

Tourism and recreation as an object of cross-border research are studied to a lesser extent. For the theoretical substantiation of this subject, the works of Zaitseva et al. (2016), Mazhar (2009), Karpova and Valeeva (2015) are of great importance. Cross-border territorial formations of tourism and recreation (systems, territories, clusters, regions) are studied by Alexandrova and Vladimirov (2016), Zyryanov (2012), Dmitrevsky (2000), Dragileva (2006), Dunets (2011), Korneevets et al. (2015), Korneevets (2010a, 2010b), Kropinova (2011, 2013, 2016), Mirzekhanova (2015), Mitrofanova (2010). The number of applied studies of cross-border cooperation along the borders of the Russian Federation is insignificant.

Defining Cross-Border Tourism and Recreation Region

A "transboundary tourism and recreation region" could be understood as compactly located territories of neighboring regions of neighboring countries that have a common tourist resource that ensures the functioning of territorial combinations of economic entities united by substantial ties (Fig. 1).

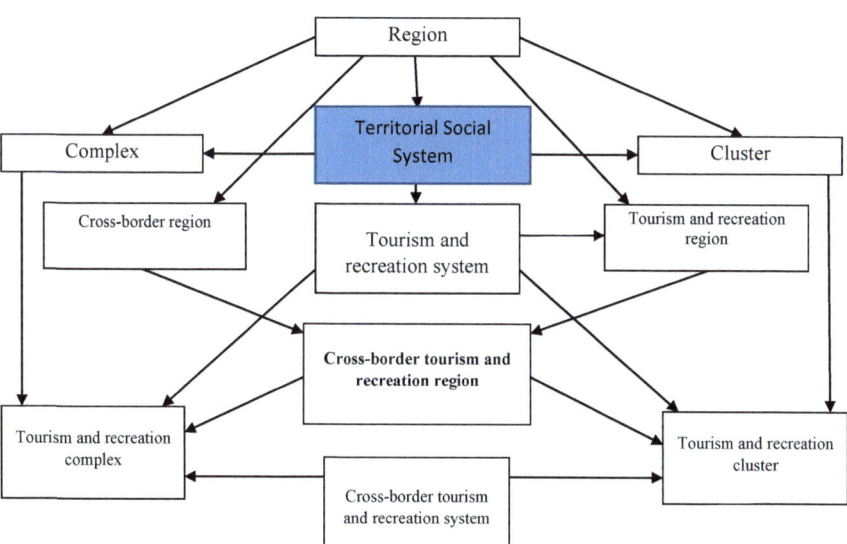

Fig. 1 Cross-border tourism and recreation regions (CTRR) in the system of notions of the social geography

The concept of a cross-border tourism and recreation region is included in the system of concepts of public geography, being, on the one hand, a functional subsystem of a cross-border region, and, on the other hand, it is a kind of tourist-recreational region.

The methodological basis of the cross-border tourism and recreation region (CTRR) research should be a systematic approach based on the consideration of the concept of CTRP as a particular notion in relation to the most general concept of the region, and the research itself as a system of complementary methods based on various regional approaches relating to different aspects of the development of the region.

Signs of Regional Formation

To determine the presence or absence of CTRP could be possible on basis of a number of features, which fall into either the region-forming category or the additional feature category.

The region-forming characteristics include:

– close links between the constituent subjects of tourism in the region;
– continuity of the territory, that is, the possibility of direct transport mode of communication, without crossing the external borders of the region;
– complementary components of the natural and cultural-historical potential of tourism development;
– availability of a common tourist resource;
– availability of a common or coordinated transport infrastructure;
– development of Internet communications;
– the availability of cross-border tourist routes;
– availability of state and public institutions coordinating cross-border tourist flows;
– signed and working cooperation agreements.

Additional region-forming features:

– similar level of socio-economic development;
– ethnic similarity of the population;
– existence of shared history;
– developed links in the social sphere;
– implementation of joint research projects.

Application of the Theory of the Formation of the Cross-Border Tourism and Recreation Regions for the Baltic Region

Transport communications developed between parts of the Baltic region play a significant region-forming role. Figure 2 shows railway and sea transport routes of the region.

All the parts of the region are connected by air-ways. Developed cross-border transport routes facilitate the use for tourism and recreation as a common natural resource—the Baltic Sea, and the other complementary components such as mountain-skiing in Scandinavia, beach and medical tourism in the South and South-Eastern Baltic, urban/intellectual tourism in capitals and other large cities, and in specially created major tourist and recreational centers (Fig. 3).

Between the subjects of tourism in the region, close ties are developing, including those flourishing through the formation of international hotel chains—Sokos, First hotels, restaurant and retail chains.

There are numerous international ferry and cruise routes on the Baltic region. The general body coordinating the development of tourism in the Baltic region is the Union of the Baltic Sea States and the Baltic Tourism Commission.

The cumulative effect of region-forming factors allows us to assert that the sectoral transboundary tourist-recreational macroregion is being constructed around the Baltic Sea.

The development of tourism in the Baltic region countries to the great extent depends on the level of socio-economic development. More developed countries, characterized by an increased level of GDP per capita production, have a higher value of the number of tourist arrivals per capita as well (Fig. 4). Several deviations of close to direct dependence could be observed for Estonia and Denmark. In Estonia, the number of arrivals is higher, while Denmark is lower than it is in the countries with a similar level of GDP per capita production.

Compiled by the author on the basis of data: Federal Service …; Eurostat; World Economic Outlook Database; International Monetary Fund; UNWTO.

Figures 4 and 5 present the results of the assessment of the level and dynamics of tourism development in the Baltic region.

Figure 6 shows the differentiation of countries in terms of the number of overnight stays per capita and their dynamics.

Both indicators are led by Sweden and Denmark, while Germany and Finland have a large number of overnight stays per capita. Poland, Lithuania and Estonia have high rates of development of international tourism. Russia, unfortunately, still does not occupy the best positions.

Even more, of course, there is the differentiation of the level of tourism development in the mesoregion section. On the whole, the west of the Baltic region still has better results than its eastern part. Therefore, in the area of the Danish Straits, a cross-border tourist-recreational region is the best formed. The lowest indicator is in Latvia, which lost a lot from the outflow of Russian tourists.

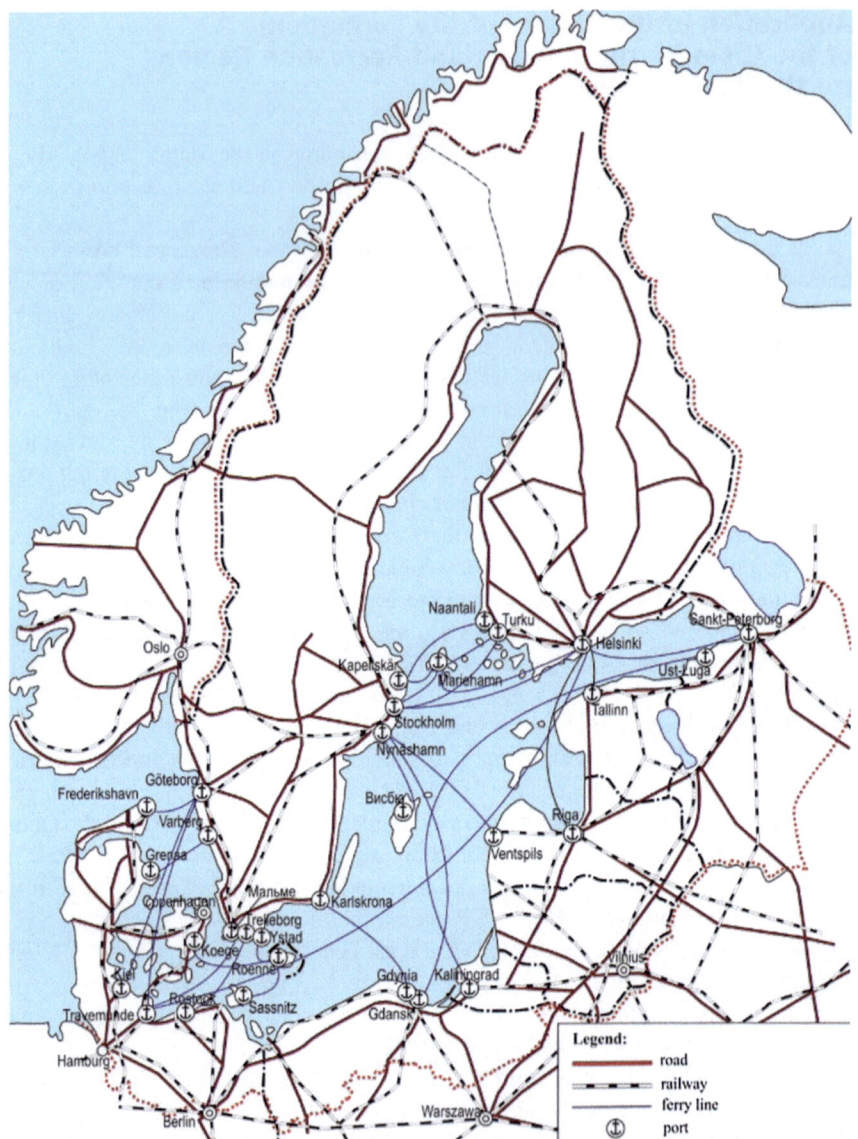

Fig. 2 The most important land and waterways that contribute to the consolidation of the Baltic region

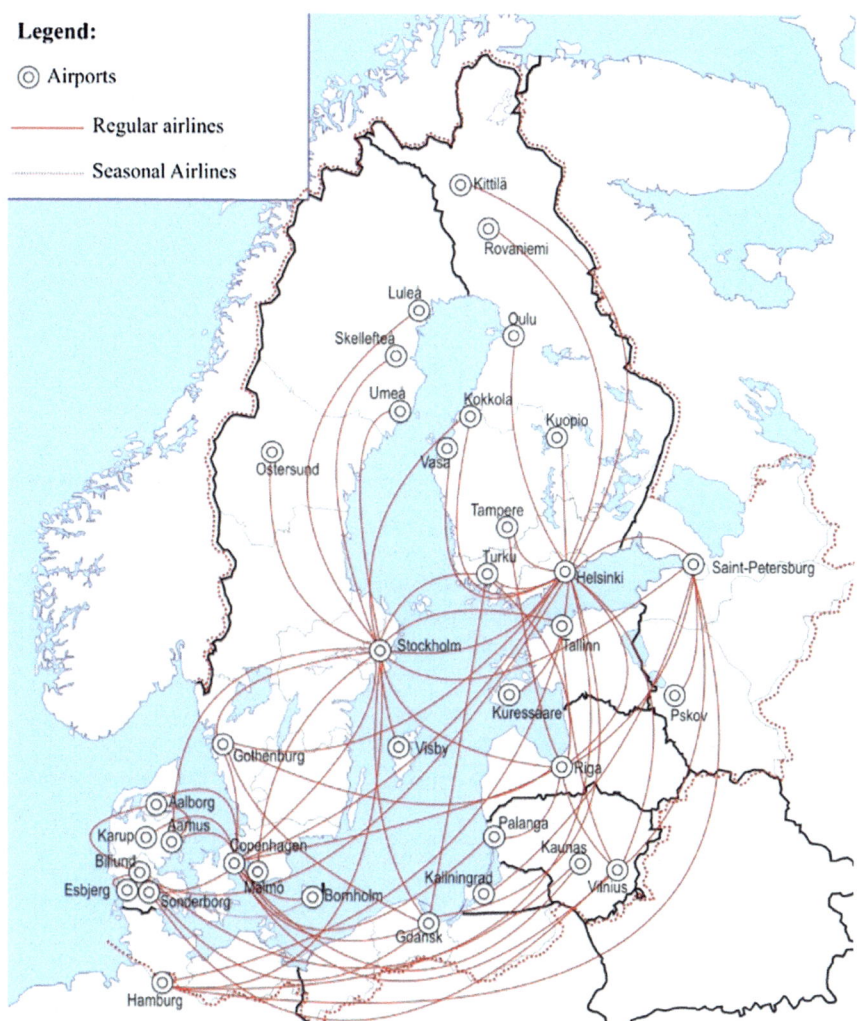

Fig. 3 Air-ports and airlines in the Baltic region

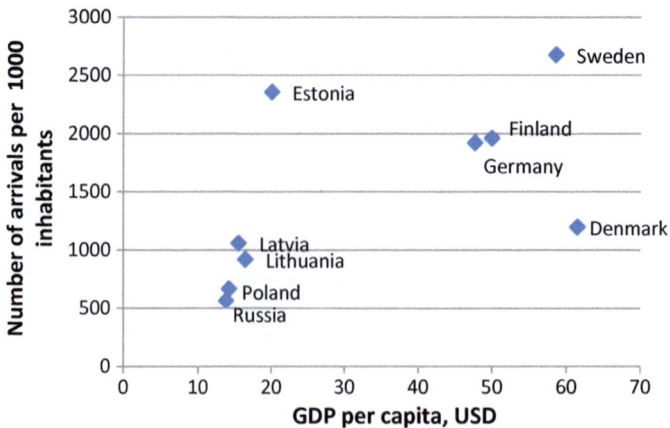

Fig. 4 Distribution of the countries of the Baltic region in the production of GDP per capita and number of arrivals per 1000 inhabitants

Figure 7 demonstrates the cross-border tourism and recreational mesoregions in the Baltic macroregion. There are at least eleven of them could be distinguished. In the activities of six of them, the subjects of the Russian Federation participate.

No. of mesoregion	Composition of mesoregions	Specialization
1	German-Danish-Swedish-Polish	Cognitive, cruise, yachting
2	Polish-Russian-Lithuanian	Shopping tourism, historical and cultural
3	Latvian-Estonian-Russian	Cultural and historical
4	Finnish-Estonian	Cruise
5	Russian-Estonian	Shopping tourism, historical and cultural
6	Russian-Finnish southern	Cruise, weekends
7	Russian-Finnish northern	Shopping tourism, nature-oriented
8	Russian-Finnish-Norwegian	Shopping tourism, cultural, business
9	Swedish-Norwegian	Skiing
10	Swedish-Norwegian	The nature-oriented
11	Swedish-Danish	Cultural-historical, weekends

Jointly, the effect of economic factors—the advantages in increasing the competitiveness of both the international and domestic markets of territories participating in cross-border cooperation, on the one hand, and the institutional basis for cooperation between neighboring regions of different countries, on the other hand, determines the formation of cross-border tourism and recreational mesoregions (Fig. 7). They are at different stages of formation, depending on the activity of the

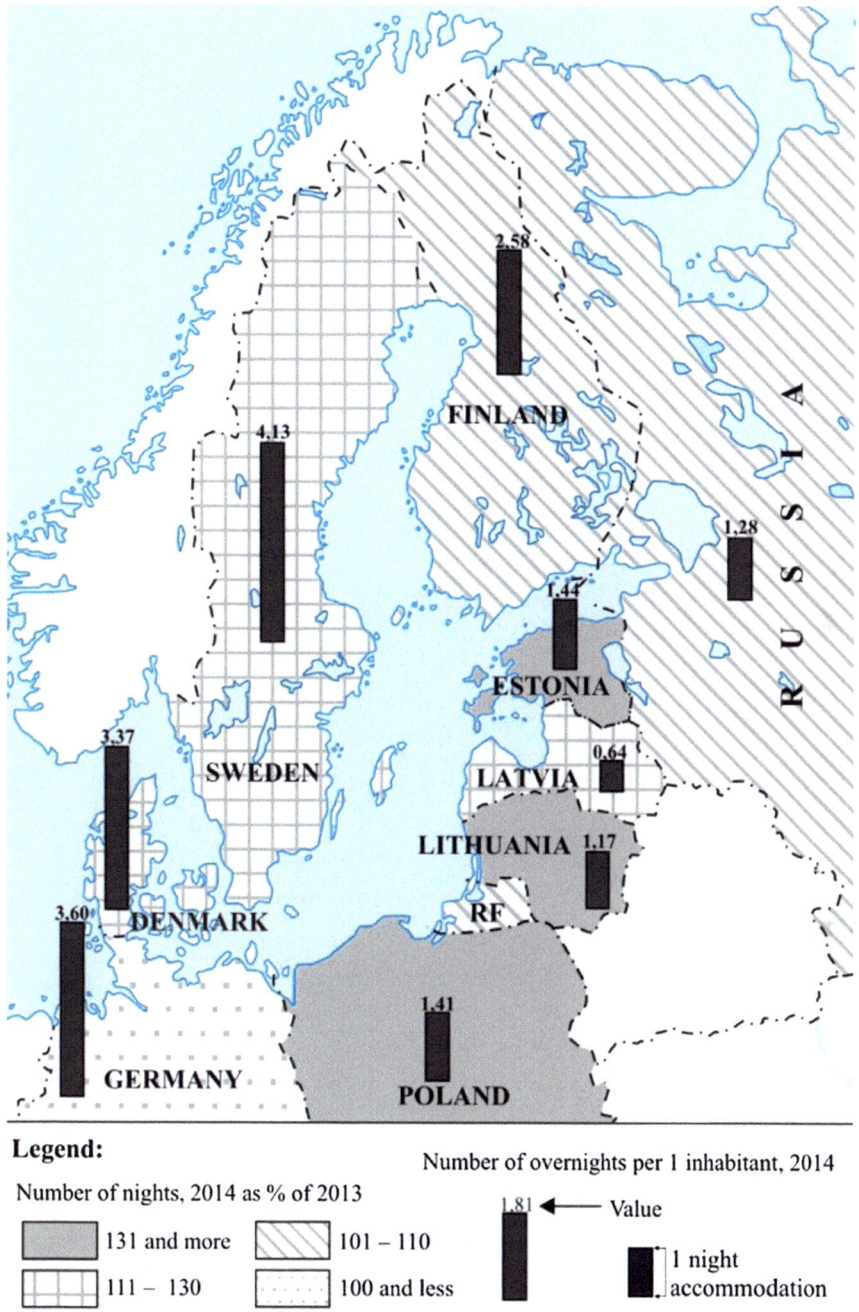

Fig. 5 Estimation of the level and dynamics of development of tourism and recreation in the countries of the Baltic region on the basis of the number and dynamics of the number nights spent (compiled by the author on the basis of data: Federal Service …; Eurostat)

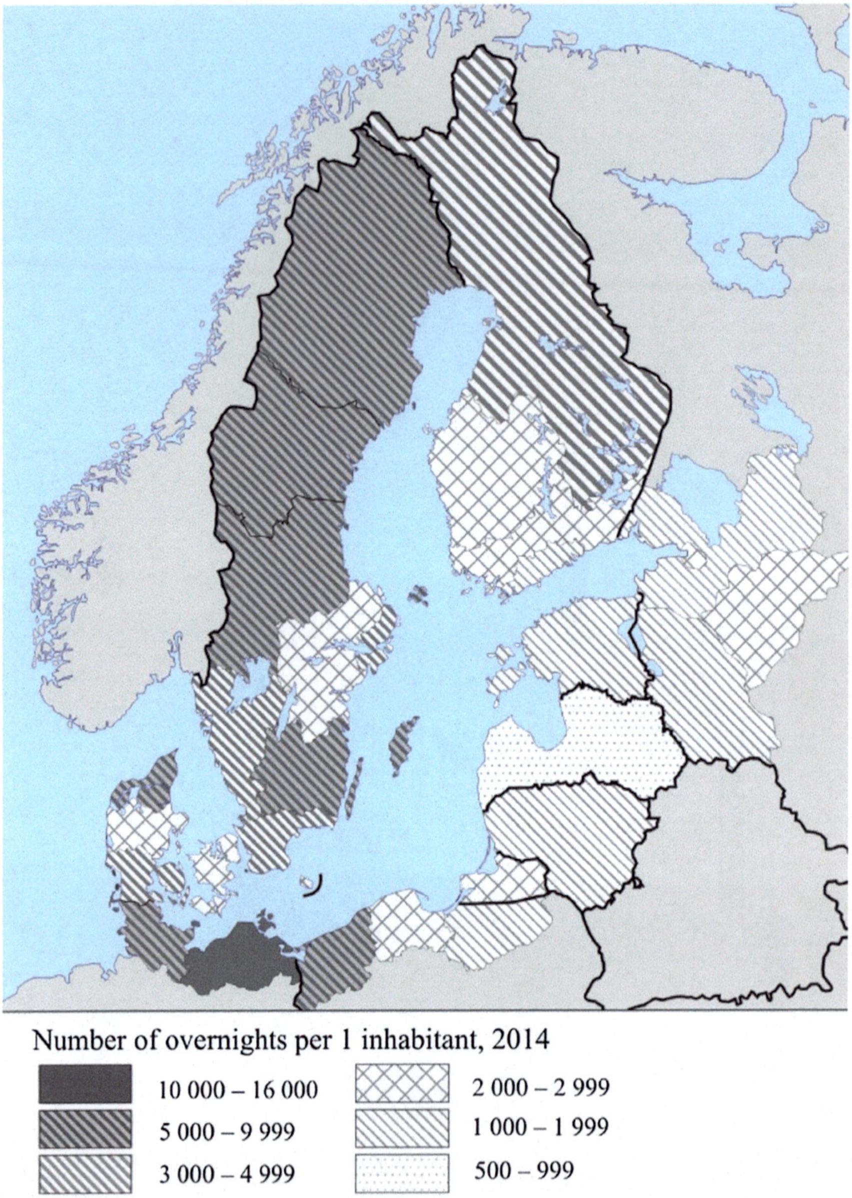

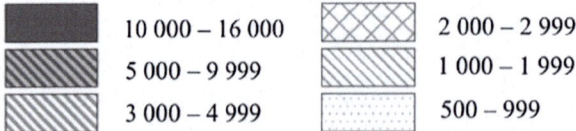

Number of overnights per 1 inhabitant, 2014

■	10 000 – 16 000	▨	2 000 – 2 999
▨	5 000 – 9 999	▧	1 000 – 1 999
▧	3 000 – 4 999	░	500 – 999

Fig. 6 Level of development of recreation and tourism in the mesoregions of the Baltic CTRR, 2014

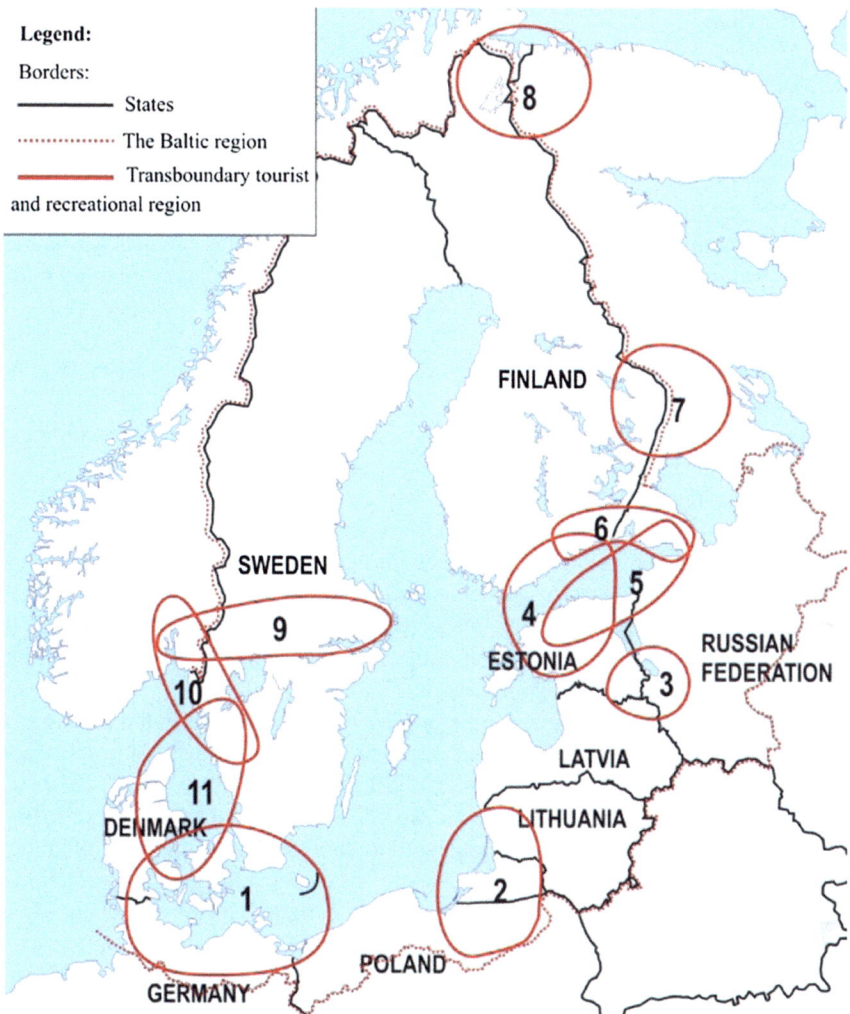

Fig. 7 Cross-border tourism and recreation mesoregions in the Baltic macroregion

cooperating regions. The most formed is the German-Danish-Swedish-Polish CTRR located on both sides of the Danish straits. There are significant tourism and recreational connections within the Finnish-Estonian mesoregion. At the initial stages of formation are the Russian-Finnish (northern) and Russian-Finnish-Norwegian CTRRs. The rest of the regions have an average level of formation.

Conclusion

In the context of globalization, the formation of cross-border, including tourist-recreational, regions is a regular occurrence. These can be formed on macro-meso- and microlevels.

Cooperation with the regions of neighboring countries and the formation of transboundary tourism and recreational regions contribute to boosting competitiveness of tourism and recreation in the border regions of the interacting countries.

The following factors play an important role in the formation of cross-border regions:

– Political: the predominance of the contact function of the boundary over the barrier;
– Economic: level of development in the countries of tourism and recreation;
– Institutional: national legislation and interstate treaties and agreements;
– Natural: as prerequisites for the joint use of natural objects.

The Baltic region is one of the most formed cross-border tourism and recreational regions with Russian participation. Here, it is necessary to monitor the processes of cross-border tourism and recreation region formation both through the implementation of national research projects and within the framework of new cross-border cooperation programmes.

Cross-border cooperation programmes play a significant role in the processes of cross-border tourism region formation. In April 2018, the first call for proposals for the Lithuania-Russia Cross Border Cooperation Programme 2014–2020 was completed. Of the 57 applications submitted, 27 were devoted to priority 1.1, "P1.1: Restoration and adaptation of historical and natural heritage, promotion of culture, cultural networking and tourism development" of the thematic goal 1 "Promotion of local culture and preservation of historical heritage" (The Lithuania-Russia Cross-border Cooperation Programme 2014–2020). There is similar proportion in distribution among the applications for the Poland-Russia Cross Border Cooperation Program 2014–2020—57 of 118 applications were submitted on the thematic objective "Heritage" (The Poland-Russia Cross-border Cooperation Programme 2014–2020).

References

Alexandrova A, Vladimirov Yu (2016) Osobennosti sozdaniya turistskikh klasterov v Rossii (na primere Vologodskoy oblasti) (Features of the creation of tourist clusters in Russia (on the example of the Vologda region). Mod Probl Serv Tour 10(1):47–58 (in Russ.)

Baklanov P (2014) Geographicheskie i geopoliticheskie factory v regionalnom razvitii (Geographical and geopolitical factors in regional development). Reg Stud 2:4–10 (in Russ.)

Baklanov P, Hansey S (2004) Prigranichnye I transgranichnye territorii kak object geographicheskikh issledovanij (Cross-border and transboundary territories as an object of geographical research). Izvest RAN Ser Geogr 4:27–34 (in Russ.)

Brym M (2013) Border landscapes on the EU periphery: examples of integration and division Polish and Ukrainian cities. Focus Geogr 56(3):84–94

Dmitrevsky Yu (2000) Turisticheskie region mira: uchebnoje posobie (Tourist regions of the world: a teaching guide). SSU, Smolensk (in Russ.)

Dragileva I (2006) Transgranichnoye sotrudnichestvo v rasvitii turisma Jugo-Vostochnoi Baltiki (Transboundary cooperation in the development of tourism in the South-Eastern Baltic). Dissertation, St. Petersburg: SPbSU (in Russ.)

Dunets A (2011) Turism v transgranichnyhk territorijah Altae-Sajanskogo gornogo regiona (Tourism in transboundary territories of the Altai-Sayanskogo Mountain Region). Reg Stud 1 (31):87–91 (in Russ.)

Eurostat. Official portal. http://ec.europa.eu/Eurostat. Accessed 20 May 2018

Federal State Statistics Service. http://www.gks.ru. Accessed 1 Aug 2017

Fedorov G (2016) Problemy transgranichnoto regionoobrazovaija v rossijskih regional na Baltike (Problems of cross-border regional formation in the Russian regions in the Baltic). Socio-economic geography. Bull Assoc Russ Soc Geogr 5:82–92 (in Russ.)

Gal Z (2009) Danube Region—past, present and future prospect of trans-national cooperation as a playground of the European Integration. Eurolimes 7:148–159. http://arsiv.setav.org/ups/dosya/17095.pdf. Accessed 17 Jan 2016

Gerasimenko T (2012) Rol etnokulturnoj osnovy v formorovanii transgranichnyh regionov (The role of the ethno-cultural basis in the formation of transboundary regions). Sociocultural regional conduct. Center for the Study of Interaction of Cultures. http://www.regionalstudies.ru/journal/homejornal/rubric/2012-11-02-22-16-38/268–q-q.html. Accessed 10 Aug 2015 (in Russ.)

Gerasimenko T, Gladkiy Y (2005) Transgranichnost kak factor etnokulturnogo I economich-eskogo razvitija (Cross-borderness as a factor of ethno-cultural and economic development). Izvest RGO 137(6):1–9 (in Russ.)

Grama V (2011) Cooperation at the EU External Borders. Case of Ramania. Roman Rev Polit Geogr 13(2):256–263

Karpova G, Valeeva Y (2015) Razvitie turizma kak konkurentnoe-preimuschestvo territorii na primere subektov SZFO (Development of tourism as a competitive advantage of the territory (on the example of the subjects of the North-West Federal District)). Izvestiya of the St. Petersburg State Economic University 1(91):45–51. http://cyberleninka.ru/article/n/razvitie-turizma-kak-konkurentnoe-preimuschestvo-territorii-na-primere-subektov-szfo. Accessed 01 Oct 2017 (in Russ.)

Katrovsky P (2015) Tendenzii I problem razvitija prigranichnyh s Belorussiej regionov Rossii (Tendencies and problems of development of the regions of Russia bordering with Byelorussia). Izvest Smolensk gos. University 5:162–173 (in Russ.)

Kolosov V, Turovsky R (1999) Tipy novyh rossijskih graniz (Types of new Russian borders). Izvest RAN Ser Geogr 5:30–47 (in Russ.)

Kolosov V, Vendina O (2011) Rossijsko-ukrainskoje pogranichie: dvadzat let rasdelennoto edinstva (Russian-Ukrainian borderland: twenty years of divided unity). M.: A new chronograph (in Russ.)

Korneevets V (2010a) Mezhdunarodnye, transnazionalnye I transgranichnye region: prisnaki, osobennosti, ierarhija (International, transnational and cross-boundary regions: signs, features, hierarchy). Vestnik of the I. Kant Russian State University 1:27–34 (in Russ.)

Korneevets V (2010b) Formirovanie transgranichnyh mesoregionov na Baltike (Formation of cross-border meso-regions at the Baltic). I. Kant Russian State University, Kaliningrad (in Russ.)

Korneevets V, Kropinova E, Dragileva I (2015) The current approaches to the cross-border studies in the sphere of tourism. Int J Econ Financ Issues 5:66–73

Kovacs A (2011) On borders. Border regions and cross-border retail-trading. Scientific Papers of the University of Pardubice. Series D. Faculty of Economics & Administration. Pardubice 28:29–42

Kropinova E (2011) Faktory formirovanija transgranichnogo turistsko-rekreazionnogo regiona "Jugo-Vostochnaja Baltika" (Factors forming the transboundary tourist-recreational region "South-Eastern Baltic"). Balt Reg 1:106–114 (in Russ.)

Kropinova E (2013) Sotrudnichestvo mezhdu Rossiej I EU v sphere innovazionnogo razvitija turisma na primere programmy prigranichnogo sotrudnichestva "Litva-Polsha-Rossia" (Cooperation between Russia and the EU in the field of innovative tourism development: case for the cross-border cooperation program "Lithuania-Poland-Russia"). Balt Reg 4(18):67–80 (in Russ.)

Kropinova E (2016) Komplexny podhod pri planirovanii I upravlenii transgranichnymi turistskimi klasterami (na primere rossijsko-litovskogo-polskogo transgranichnogo turistsko-rekreazionnogo regiona, prilegajuzhego k ozeru Vizhtynezkoje) (An integrated approach to the planning and management of a transboundary tourism cluster (The case of Russian-Lithuanian-Polish cross-border tourism and recreational region, adjacent to the Vishtynetskoe Lake). Mod Probl Serv Tour 10(1):117–128 (in Russ.)

Lagiewski R, Revelas D (2004) Challenges in cross-border tourism regions. Rochester Institute of Technology: RIT Scholar Works. http://scholarworks.rit.edu/cgi/viewcontent.cgi?article=1554&context=other. Accessed 10 October 2016

Manakov A (2014) Dinamika transgranichnyh kontaktov naselenija zapadnyh rajonov Pskovskoj oblasti v 2003–2013 godah (Dynamics of cross-border contacts of the population of the western regions of the Pskov region in 2003–2013). Pskov Reg J 17:98–107 (in Russ.)

Mazhar L (2009) Territorialnye turistsko-rekreazionnye systemy: geosystimny podhod k formirovaniju I razvitiju (Territorial tourist-recreational systems: geosystemic approach to formation and development). Dissertation, Smolensk (in Russ.)

Mezhevich N, Kretinin G, Fedorov G (2016) Economic and geographical structure of the baltic sea region. Balt Reg 8(3):11–21. https://elibrary.ru/item.asp?id=27446394.

Mirzekhanova D (2015) Klasternyi podhod v rasvitii turizma (na primere Khabarovskogo kraja) (Cluster approach in the development of tourism: case for the Khabarovsk Territory). News Altai State Univ 4:87–95 (in Russ.)

Mitrofanova A (2010) Regionalny turistskyi kaster kak forma rostranstvennojy organisazii turisma (na primere Kaliningradskoj oblasti) (Regional tourist cluster as a form of spatial organization of tourism: case for the example of the Kaliningrad region). Dissertation, I. Kant Russian State University (in Russ.)

Prokkola E-K (2008) Resources and barriers in tourism development: cross-border cooperation, regionalization and destination building at the Finnish-Swedish border. Fennia 186(1):31–46. http://ojs.tsv.fi/index.php/fennia/article/viewFile/3710/3500. Accessed 20 Sept 2016

Scott J, Laine J (2012) Borderwork: Finnish-Russian co-operation and civil society engagement in the social economy of transformation. Enterp Reg Dev 24(3–4):187–197

Seric N, Marcovic S (2011) Brand management in the practice of cross-border tourism destinations. Acad Tur 2:89–99

Studzieniecki T, Palmowski T, Korneevets V (2016) The system of cross-border tourism in the Polish-Russian borderland. Proc Econ Financ 39:545–552

The Lithuania–Russia Cross-border Cooperation Programme 2014–2020. http://www.eni-cbc.eu/lr/en/eng. Accessed 20 Aug 2018

The Poland–Russia Cross-border Cooperation Programme 2014–2020. https://www.plru.eu/en/news. Accessed 20 Aug 2018

Timothy D (1999) Cross-border partnership in tourism resource management: international parks along the US–Canada border. J Sustain Tour 3–4:182–205

Turovsky R (2004) Strukturny, landshaftny I dinamicheskit podhody v kulturnoj geographii. (Structural, landscape and dynamic approaches in cultural geography). In: Humanitarian geography, vol 1. Institute of Heritage, Moscow (in Russ.)

Vardomskiy L (2017) Post-Soviet integration and economic growth of the new borderland of Russia in 2005–2015. Prostranstvennaya Ekonomika = Spatial Economics, 2017, no 4, pp 23–40. https://doi.org/10.14530/se.2017.4.023-040 (in Russ.)

Vodeb K (2010) Cross-border regions as potential tourist destinations along the Slovene-croatian frontier. Tour Hosp Manag 16(2):219–228

Wachowiak H (2006) Tourism and borders. Aldershot, England. VT Ashgate, Burlington

World Economic Outlook Database. https://www.imf.org/external/pubs/ft/weo/2018/01/weodata/index.aspx. Accessed 20 Aug 2018

Zaitseva N, Korneevets V, Kropinova E, Kuznetsova T, Semenova L (2016) Cross-border movement of people between Russia and Poland and their influence on the economy of border regions. Int J Econ Financ Issues 6(4):1690–1695

Zyryanov A (2012) Geographicheskoye pole turistskogo klastera (Geographical field of the tourist cluster). Geogr Bull 1(20):96–98 (in Russ.)

Cross-Border Clustering Across the Baltic Region: Relating Smart Specialization and Cluster Categories

Andrey S. Mikhaylov

Abstract

Clusters are complex cross-sectoral and inter-organizational formations stretching beyond the statistical limitations of industry classifications. Cluster mapping techniques often rely on core industries in defining core products or technologies acting as nucleus for bringing together the heterogeneity of economic entities. Awareness of non-synonymy between the notion of cluster and industry are essential in efficient implementation of regional development policies. Internationalization and cross-border integration of regional industries further complicates the delimitation process of clusters and brings up new challenges for smart specialization strategies. The study stresses on particular features of cross-border clusters and raises a non-trivial discussion over international division of labor in the context of regional smart specialization.

Keywords

Baltic region · Cross-border cooperation · Cross-border cluster ·
Smart specialization · Cross-sectoral cluster · International division of labor

Introduction

Regions are complex spaces differentiated by certain properties and qualities—environmental, institutional, socio-economic, cultural, historical, political, etc. Most particularities are being formed over a long period of time, with natural factors (e.g. climatic, geological, etc.) of the ecosystem determining the basic development trajectory of local societies. Some regional communities exhibit strong similarities,

A. S. Mikhaylov (✉)
Immanuel Kant Baltic Federal University, Kaliningrad, Russia
e-mail: mikhailov.andrey@yahoo.com

© Springer Nature Switzerland AG 2020
G. Fedorov et al. (eds.), *Baltic Region—The Region of Cooperation*,
Springer Proceedings in Earth and Environmental Sciences,
https://doi.org/10.1007/978-3-030-14519-4_11

while others are completely different. Spatial proximity of borderland regions implies an increased probability of having identical natural environment, intensified social relations (including historical shifts in borderline), and even common institutional setup at cross-border or transnational level (e.g. Euroregions, European Union, etc.). At its extreme national border regions may have more in common with adjacent territories of the neighboring state than within a nation-state (causing preconditions to voluntary accession or even annexation) or a significant divergence from both, which might trigger referendum on sovereignty. Thus, inter-regional proximity factors define the intensity and scope of cross-border cooperation and integration of the two (or more) territorial socio-economic systems.

Cross-border cooperation is an ordinary process for resembling border regions. Often speaking the same language (e.g. the French-speaking region of Romandy in western Switzerland and the adjacent territories of France), dominated by same religious denomination, having close family ties, intensified population mobility and other linkages, these regions have sustainable historically grounded socio-economic relations. The narrow gap in intellectual, institutional, organizational, socio-cultural, technological setting of the regional community (incl. households, business, academia, etc.) makes networking easy, facilitating inter-organizational networking across borders (incl. informal knowledge dissemination—the 'local buzz' effect; Bathelt et al. 2004). Some scholars suggest that the innovation efficiency of these linkages is limited due to the little divergence in the knowledge base (Boschma 2005; Mattes 2012). However, it provides an increased absorptive capacity that extends the opportunity for utilizing synergies across industries.

The overwhelming majority of border regions in the Baltic region have initiated cross-border initiatives for boosting entrepreneurial cooperation (Kern and Löffelsend 2004; Mikhaylov and Mikhaylova 2014; Pacuk et al. 2018; Pikner 2008; Scott 2002). On average a quarter of all international cooperation projects are focused on supporting business activity over the border (Mikhailov 2014). Regions are active in engaging complementary stakeholders from adjacent territories of neighboring states. The cross-border regional clusters are being established setting the benchmark for all borderland territories.

The aim of the article is to assess the complexity of new cluster mapping techniques based on cluster categories and to discuss policy implication with regard to regional specialization strategies in cross-border regions. The paper proceeds with the review on conceptual grounds of spatial networking across borders and economy sectors. Section 3 gives an overview of the research methodology applied. The research findings are presented in Sect. 4. The article concludes with discussion on managerial implications of regional smart specialization.

Literature Review

Smart specialization implies a thought-through consolidation of resources on the major domains of regional territorial capital. With that, it is clear that regional industrial, educational, innovation, etc. policies should transcend their limited focus area as to take into consideration not only the local nodes of competitiveness, but also the involvement in extra-regional networks of knowledge generation and exploitation (Chen 2015; Holl and Rama 2009; Sternberg 2007). Assessment of external synergies is particularly important for borderland regions, whose objective is to capitalize on regional strength and exercise measures for leveling out the impact of existing weaknesses (e.g. lack of individual infrastructure units, inconvenient logistics, raw-resources shortage, insufficient competences in a particular area, etc.) by integrating resources across borders.

For most borderland regions being the geoeconomic periphery of their countries, spatial proximity to an additional source of input is a significant competitive advantage. Knowledge is herewith the predominant input for boosting the innovation activity in the region. Locally established knowledge base is subjected to path-dependency for all institutional helices of the regional innovation system (Lagerholm and Malmberg 2009; Martin and Sunley 2006). Universities tend to be immobile with regard to their existing curriculum, with faculty adhering to established research areas (e.g. the focus tends to remain as new postgraduate students undertake their research on similar topics). Business displays a profound technological lock-in effect, as described by numerous scholars (Cowan and Hultén 1996; Uotila et al. 2017; Østergaard and Park 2015). The implementation of change in production process is complicated not only by need for investments, but the opposition of personnel as well, including the top management. Non-governmental organizations (NGOs) often represent the standpoint of low and middle-level staff being highly vulnerable to any radical shift in required competences and capabilities by the local labor market. Public authorities adhere to long-term development strategy, balancing between social tranquility and accelerated development. Therefore, reinforcement of cross-border contacts of regional communities is critical to sustaining innovation activity. This statement is confirmed by many researches, including those featuring studies on the western borderland of Russia (Druzhinin 2008; Fedorov 2010).

Cross-sectoral knowledge spillovers are the major source of radical innovation. Studies held on spatial-networking of related industries (e.g. by inputs, technologies, markets, etc.) reveal an increased performance of the actors involved (Delgado et al. 2016; Gaschet et al. 2017). Cluster mapping studies confirm the allegations on cluster policies made by Lindqvist et al. (2003), who advocate for inconsistence of industry classifications to regional clusters and objectify the need to differentiate between industrial and cluster policies. Numerous in-depth case studies further broaden the perception of clusters as inter-industry value chains by incorporating untraded relations with heterogeneous entities representing all institutional helices of the region—academia, government, NGOs, and society.

The broad range of actors engaged in a value co-creation process is accompanied by a wide spectrum of market offerings. Each given entity is a participant of numerous relations across networks and industries. Following a particular subject-specific activity (generally focused on industry sector) the services rendered and goods delivered can be ranked from primary to secondary ones. Therefore, the networking stakeholders may be defined as being of major or minor importance for a given cluster specialization. This sets a prefiguration of concentric waves from conditionally the major actors of the networking—whose disappearance will lead to the collapse of a given network. For example, vineyards are primary for California wine cluster, while wine educational centers are secondary. This logic forms the basis for the new cluster mapping techniques that rely on so called cluster categories (or definitions). As suggested by Delgado et al. (2016, p. 1), these cluster categories represent "groups of industries closely related by skill, technology, supply, demand, and/or other linkages", thus, input-output tables or a supply chain is only part of existing interrelations (although being a great importance). Therefore, the main obscurity lies in the set of statistically defined indicators to be considered, being especially subjected to significant methodological limitations when considering an international configuration of clusters.

Methodology

The narrow connotation of the Baltic macro-region limits this area to countries and regions having direct access to the Baltic Sea. These are ten European countries, including Russia (the Kaliningrad and Leningrad regions), the Nordic countries (Denmark, Finland, and Sweden), the northern regions of Germany and Poland, and the Baltic States (Estonia, Latvia, and Lithuania). There are 18 borderland regions corresponding to second level of a common classification of territorial units for statistics (NUTS 2) of the European Commission (2015). A long-standing strategy for cross-border cooperation in the Baltic region has resulted in numerous Euroregions formed, some of which are among the first ones to be created—North Calotte Council, 1971 (Finland, Norway, Sweden), Kvarken council, 1972 (Finland, Norway, Sweden), Central North committee, 1977 (Finland, Norway, Sweden). Yearly studies suggest that Scandinavian countries are well ahead in creating a common socio-economic space (the Nordic Committee for Economic Cooperation est. in 1948, the Nordic Council est. in 1951), giving rise to striking (although not ideal) cross-border cooperation projects (e.g. Oresund cross-border region project; Hall 2008; Hospers 2006; Schmidt 2005). Of particular interest in this study is the cross-border area of Oresund region (Sweden) and Greater Copenhagen (Denmark). The Euroregion was officially established in 2000 with the opening of the Oresund Bridge connecting the two countries, giving rise to intensive cooperation across borders.

The Oresund cross-border region hosts widely recognized cross-border clusters of different specialization—life sciences, information technology, food processing etc. The most prominent is the Medicon Valley cluster of life science (see: www. mediconvalley.com). Being led by companies specialized in medical and biotechnology the cluster organization unites over 200 companies specialized in medical technology, 155 biotech organizations, 80 contract research organizations and other institutions with an overall employment reaching 40,000 people (Medicon Valley: facts and figures 018). Most member organizations are densely clustered around

Copenhagen—the capital city of Denmark, and the two major cities of the Scania province—Lund and Malmo (see: www.mediconvalley.com/industry)

The Medicon Valley has become a case of good practice for cross-border cluster initiatives and a benchmark for validating hypothesis and assumptions. Availability of qualitative and quantitative data on the cluster network as well as the long established reputation for one of the few de facto exiting cross-border clusters makes it the perfect sample for testing the predictive capacity of statistical indicators for the cross-sectoral cluster categories. The initial matrix of cluster categories applied is a combination of the originally presented scheme of related clusters developed by the Institute for Strategy and Competitiveness of the Harvard Business School in partnership with the U.S. Department of Commerce and U.S. Economic Development Administration (see: www.clustermapping.us/cluster) and the European Cluster Panorama 2016 project of the European Cluster Observatory (Ketels and Protsiv 2016). European approach has alienated the medical devices industry into a separate category, providing its unfolded categorization, whereas initial (i.e. American) cluster category merges the two groups. Current methodology implies cumulative account of the two cluster categories, while considering the detailed picture given by European Cluster Panorama. Further analysis is based on an assumption of two major (core) and four minor (supporting) interrelated industries in terms of Medicon Valley life science cluster. These are: (a) BIO: (1) Biopharmaceuticals; (1.1) Upstream Chemical Products; (1.2) Downstream Chemical Products; and (b) MED: (2) Medical Devices; (2.1) IT and analytical instruments; (2.2) Lighting and electrical equipment.

The source of industry-specific data is the European cluster observatory statistical data presented at the European Commission website (see: www.ec.europa.eu/growth/smes/cluster/observatory/cluster-mapping-services). The study applies the following industry-sensitive indicators for a period of 2008-2013: (1) Number of employees in full time equivalent units, (2) Number of enterprises, (3) Specialization (calculated as location quotient—the ratio of total employment in a given region to the industry's share of total employment in all EU countries considered). The statistical data on regional employment and on total number of enterprises distributed in border regions of Oresund are sourced from Eurostat—the statistical office of the European Union, for 2008–2013 (see: www.ec.europa.eu/eurostat/data/database).

Research Results

The general overview of the Oresund cross-border regional statistics corresponding to the life sciences cluster categories is presented in Fig. 1.

Figures on location quotient (specialization) suggest that Hovedstaden is highly focused on biopharmaceuticals (higher than any other region of the EU), and features a significant number of personnel engaged in all inter-related sectors of life sciences. The bordering region of southern Sweden—Sydsverige, is clearly focused on medical devices with figures being nearly twice the values of the EU average.

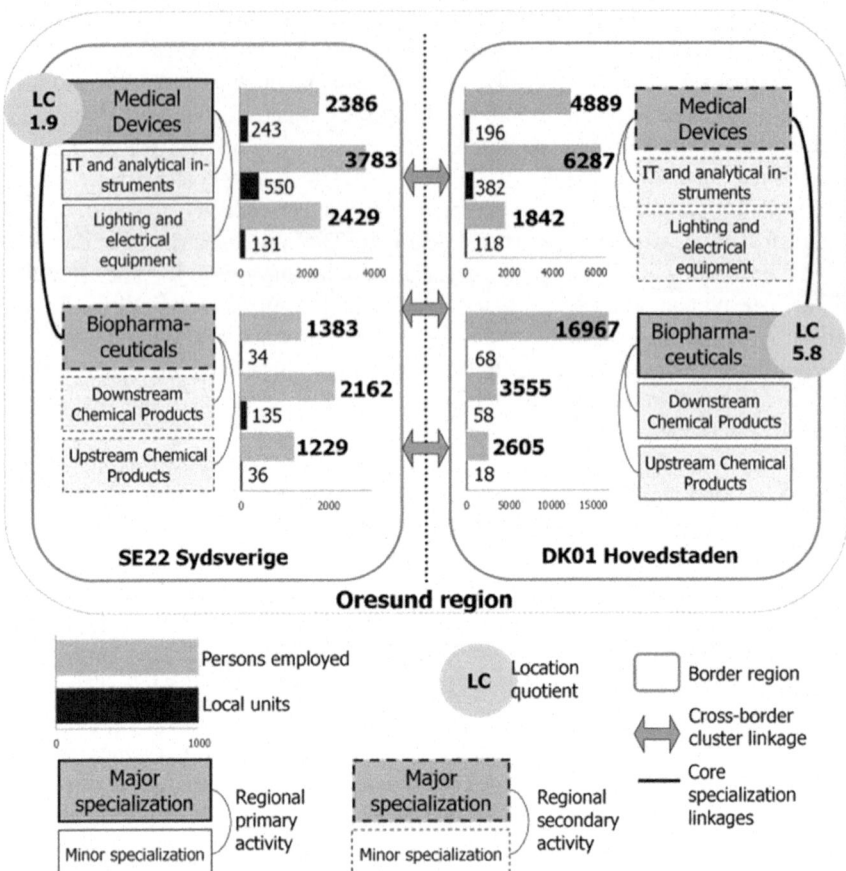

Fig. 1 The Oresund cross-border regional statistics corresponding to the life sciences cluster categories

Distribution of data across regions by core industries of the Medicon Valley life sciences cluster—biopharmaceuticals and medical devices, reveals strong polarization towards Hovedstaden—the capital region of Denmark.

The life science (biomed) cluster category employs nearly 50,000 people, which is 5% of the total employment of the cross-border region. The spatial distribution of employment figures (Fig. 2) shows that 70% (36.1 thousand people) are employed in Hovedstaden and 30% Sydsverige (13.3 thousand people).

This is despite having a greater number of companies on the Swedish side of the border. Over the past 6 years from 2008 to 2013 the share of persons employed in the cluster category decreased by 8.4% (or by 4568 people). The employment decline is found in both cluster definitions—MED and BIO, with medical devices

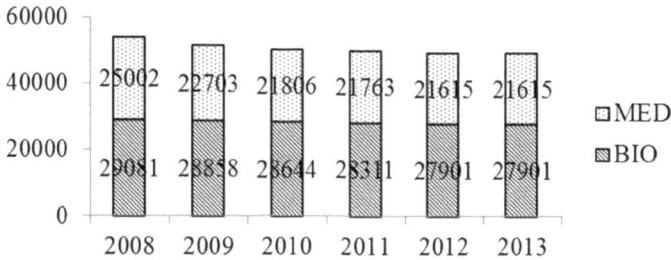

Fig. 2 Oresund region employment dynamics by specializations, people

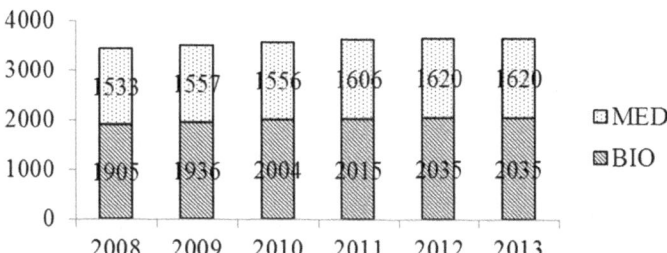

Fig. 3 Oresund region enterprise dynamics by specializations, units

featuring higher rates: 13.5 and 4.1% respectively. This tendency is unfolding against the background of the growing share of companies by 6.3% in all interrelated industries (Fig. 3).

The distribution of employed between BIO and MED major specializations is fairly equal with a slight preponderance towards medical devices by the proportion of employed and the number of companies (Fig. 4a, b).

The BIO specialization has the main reduction of personnel is in the secondary (accompanying) activity of downstream chemical products. In the MED, the main reduction in staff is found to be in primary activity of medical devices, and the interrelated secondary activity of IT and analytical instruments.

Over 1000 people were released in each of the interrelated industries. The only exception is BIO category of the Hovedstaden region, the Biopharmaceuticals и Upstream Chemical Products industries in particular. The BIO and MED categories have structural differences. For the BIO category the main concentration of people are engaged in the cluster-core activity—the Biopharmaceuticals (over 60% of employed and almost 90% of companies). For the MED category the cluster-core activity account for only 30% of companies and employees.

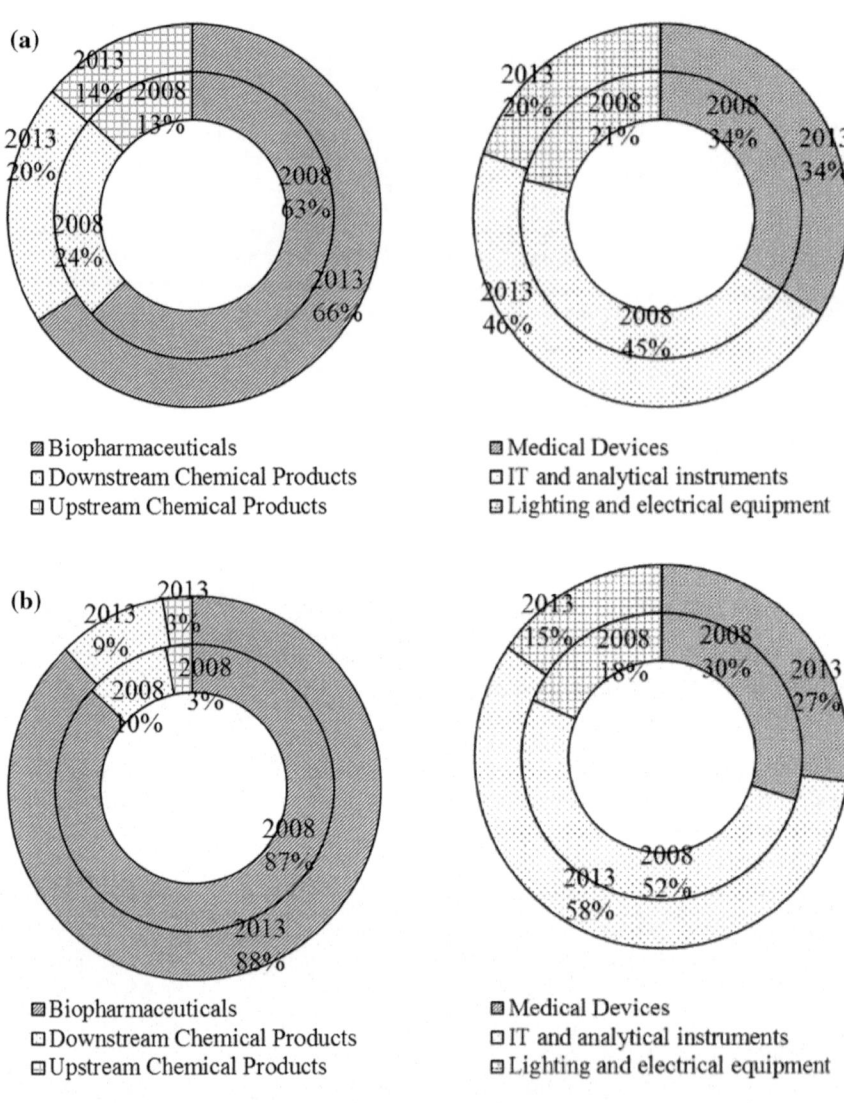

Fig. 4 **a** Oresund region employment distribution by specializations, share. **b** Oresund region enterprise distribution by specializations, share

Conclusion

Most border regions are subject to cohesion policies, both within and beyond the European Union. Regional convergence initiatives cover the social, economic, cultural, institutional and other domains, bridging the two regional systems. The intensified cross-border cooperation deepens the integrity of production networks, create cross-sectoral clusters across national borders, and trigger the establishment of a single cross-border regional innovation system. The great public concernment in such cross-fertilization and synergies results in formation of cross-border regions supported by implementation of bilateral agreements, norms and regulations, establishment of a coherent institutional context. The growing trend for international (cross-border) division of labor imposes certain specifics of cluster mapping techniques and smart specialization policies. This is an especially prominent feature for borderland regions. The interdependence and complementarity of the industries across borders suggests that structural holes might occur in regional economy structure. Individual industries might be perceived as marginal or fractional, thus, not deserving individual attention in terms of state support. Identification and acknowledgment of secondary (minor) industries in value co-creation process enables to capture the regional specifics of established (or even potential) cluster categories, raising the effectiveness of regional smart specialization policy. Economic and innovation security should come to the forefront of smart specialization practice as regional competitiveness might shift to the adjacent region. This scenario is probable is case of general socio-economic divergence of regions or an inconsistency of regional development priorities, giving an asymmetrical preference on an industry level (e.g. to a certain extend this is found to be in horticulture sector at the Limburg region of Netherlands—the North Rhine-Westphalia borderland region of Germany). The case of the Medicon Valley cluster proves the effectiveness of the cluster mapping approach based on cluster categories, which provides accurate results on inter-industry linkages.

References

Bathelt H, Malmberg A, Maskell P (2004) Clusters and knowledge: local buzz, global pipelines and the process of knowledge creation. Prog Hum Geogr 28(1):31–56. https://doi.org/10.1191/0309132504ph469oa

Boschma RA (2005) Proximity and innovation: a critical assessment. Reg Stud 39(1):61–74. https://doi.org/10.1080/0034340052000320887

Chen L-C (2015) Building extra-regional networks for regional innovation systems: Taiwan's machine tool industry in China. Technol Forecast Soc Chang 100:107–117. https://doi.org/10.1016/j.techfore.2015.07.014

Cowan R, Hultén S (1996) Escaping lock-in: the case of the electric vehicle. Technol Forecast Soc Chang 53(1):61–79. https://doi.org/10.1016/0040-1625(96)00059-5

Delgado M, Porter ME, Stern S (2016) Defining clusters of related industries. J Econ Geogr 16(1):1–38. https://doi.org/10.1093/jeg/lbv017

Druzhinin AG (2008) South Russia: integrational priorities in the big black sea region space. Sci Thought Cauc 3(55):23–30

European Commission (2015) Regions in the European Union: nomenclature of territorial units for statistics NUTS 2013/EU-28. Publications Office of the European Union, Luxembourg. http://ec.europa.eu/eurostat/documents/3859598/6948381/KS-GQ-14-006-EN-N.pdf. Accessed 21 July 2018

Fedorov GM (2010) The Kaliningrad dilemma: a development corridor or a double periphery? The geopolitical factor of the development of the Russian exclave on the Baltic Sea. Baltic Region 2(4):4–12

Gaschet F, Becue M, Bouaroudj V, Flamand M, Meunie A, Pouyanne G, Talbot D (2017) Related variety and the dynamics of European photonic clusters. Eur Plan Stud 25(8):1292–1315. https://doi.org/10.1080/09654313.2017.1306027

Hall P (2008) Policy debates: opportunities for democracy in cross-border regions? Lessons from the Øresund Region. Reg Stud 42(3):423–435. https://doi.org/10.1080/00343400701281592

Holl A, Rama R (2009) The spatial patterns of networks, hierarchies and subsidiaries. Eur Plan Stud 17(9):1261–1281. https://doi.org/10.1080/09654310903053414

Hospers G-J (2006) Borders, bridges and branding: the transformation of the Øresund region into an imagined space. Eur Plan Stud 14(8):1015–1033. https://doi.org/10.1080/09654310600852340

Kern K, Löffelsend T (2004) Sustainable development in the Baltic Sea Region. Governance beyond the nation state. Local Environ 9(5):451–467. https://doi.org/10.1080/1354983042000255351

Ketels C, Protsiv S (2016) European Cluster Panorama 2016. Stockholm School of Economics. https://www.researchgate.net/publication/320372319. Accessed 08 Aug 2018

Lagerholm M, Malmberg A (2009) Path dependence in economic geography. In: Magnusson L, Ottosson J (eds) The evolution of path dependence. Uppsala University, Sweden

Lindqvist G, Sölvell Ö, Ketels CHM (2003) The cluster initiative greenbook. Ivory Tower, Stockholm

Martin R, Sunley P (2006) Path dependence and regional economic evolution. J Econ Geogr 6(4): 395–437. https://doi.org/10.1093/jeg/lbl012

Mattes J (2012) Dimensions of proximity and knowledge bases: innovation between spatial and non-spatial factors. Reg Stud 46(8):1085–1099. https://doi.org/10.1080/00343404.2011.552493

Medicon Valley: facts and figures, 2018. http://www.mediconvalley.com/about-medicon-valley/facts-and-figures. Accessed 02 Sept 2018

Mikhailov AS (2014) Geography of international clusters in the Baltic region. Balt Reg 1(19): 113–123

Mikhaylov AS, Mikhaylova AA (2014) Spatial and sectoral distribution of international clusters in the baltic region. Eur J Sci Res 121(2):122–137

Østergaard CR, Park E (2015) What makes clusters decline? A study on disruption and evolution of a high-tech cluster in Denmark. Reg Stud 49(5):834–849. https://doi.org/10.1080/00343404.2015.1015975

Pacuk M, Palmowski T, Tarkowski M (2018) The emergence of Baltic Europe: an overview of Polish research on regional integration. Quaest Geogr 37(2):47–60. https://doi.org/10.2478/quageo-2018-0013

Pikner T (2008) Reorganizing cross-border governance capacity: the case of the Helsinki-Tallinn Euregio. Eur Urban Reg Stud 15(3):211–227. https://doi.org/10.1177/0969776408090414

Schmidt TD (2005) Cross-border regional enlargement in Øresund. GeoJournal 64(3):249–258. https://doi.org/10.1007/s10708-006-6874-5

Scott JW (2002) Cross-border governance in the Baltic Sea Region. Reg Fed Stud 12(4):135–153. https://doi.org/10.1080/714004777

Sternberg R (2007) Entrepreneurship, proximity and regional innovation systems. Tijdschr Econ Soc Geogr 98(5):652–666. https://doi.org/10.1111/j.1467-9663.2007.00431.x

Uotila J, Keil T, Maula M (2017) Supply-side network effects and the development of information technology standards. MIS Q Manag Inf Syst 41(4):1207–1226. https://doi.org/10.25300/MISQ/2017/41.4.09

Contemporary Applied Geographic Studies in the Baltic Sea Region: Geoecology

From Nature Conservation to Sustainable Development

Eugene V. Krasnov

Abstract

This paper presents the author's vision about some essential results of many years international cooperation in the environmental sciences and practice development in the Baltic countries and regions, including Kaliningrad District of Russian Federation. First interest would be given to a long-term and very successful International cooperative projects, integrated in "The Baltic University Programme" (BUP), which was coordinated by Uppsala University, Sweden. Short description of some important series of courses, which was formed by participants, would be presented. For Would Climate Research Programme was organized project "Baltic Sea Experiment" with our participations and presentation of papers about climate change in Kaliningrad District and their consequences for agriculture. The study also pays special attention to annual German–Russian Days in Kaliningrad with connection to international programme "Man and Biosphere" and development of biosphere reserves between Russia, Lithuania, and Poland. Finally, author suggest some tasks for students about modernization traditional curriculum schemes.

Keywords

Baltic University Programme · Environmental science · Sustainable region · Protected areas spatial zoning

E. V. Krasnov (✉)
Department of Geography, Environmental Management and Spatial Development, Immanuel Kant Baltic Federal University, Kaliningrad, Russia
e-mail: ecogeography@rambler.ru

Introduction

This is a brief overview of the ideas, approaches and results in the field of environmental research and education, that were achieved by the participants of the multi-year forum called "The Baltic University Programme" and some others. It is intended primarily for students, interested in methodological aspects of the development of educational technologies in close connection with research in the field of regional development. To this end, the author offers readers the task of self-comparative study and content-analysis of the main standard courses and peer-reviewed publications on the most actual ecological problems of the Baltic region and many others.

These two counter-flows of information should be compared not only as verbal models, but also in a meaningful sense, in view of the interdisciplinary approach, that brought together University professors and researchers from the Baltic region in search of better understanding and cooperation. The main purpose of this unusual task, presented in the title of the article, is to up the activity of students in relation to their choice of the most preferred cluster of formal disciplines and conscious direction of research bringing us all closer to a more sustainable development.

The Baltic University as an International Cooperative Programme

Really, this is a unique educational programme aiming at sustainable development which was generated in 1991 and realized until today by intercultural communication between students and researchers networks of 232 national universities (Ryden et al. 2003).

The Environmental Sciences courses as multidisciplinary overviews of the geosituation in the Baltic Sea region (see also Krasnov et al. 2013; Barinova and Kokhanovskaya 2015; Koroleva and Revunkov 2017; Koroleva et al. 2019) and on how to manage and protect it. The courses for undergraduate level provide an excellent background for studying the courses on sustainable development. The course can be divided into two equivalent modules:

The Baltic Sea Environment
Basic Environmental Science

which can be studied separately (Lindroos 2013).

The Baltic Sea Region Area Studies

The courses treat the region and its societies under eight headings: history, culture, language, democracy, multiculturality, social conditions, economics and security in connection with climate change. The courses deal mostly with the conditions in the region after the systems change in 1989–91. Regional development is in focus with emphasis on legal base and economics (Krasnov 1993, 1994; Kropinova 2009; Krasnov et al. 2014; Krasnov and Kropinova 2017).

Peoples of the Baltic
Regional development

The course on Sustainable Baltic Region deals with sustainable use and management of natural resources and long-term protection of the environment (Zaporozhsky and Krasnov 1996). Important issues in the course are energy and energy use, material flows, economy and ethics, industry, agriculture, transport and community development.

One very original course on "Sustainability Applied in International learning Sail", which was organized by the Baltic University Programme consists of workshops, seminars and group discussions on board a sailing ship "Fryderyk Chopin" (17–24 May 2013) from Gdynia, Poland to Visby, Sweden, and finished next in Gdynia. Participating teachers represents different universities from 12 countries. The course was focused on sustainability issues in the Baltic Sea Regions, with the aim to give a more widened understanding of the problems and possibilities for more sustainable development in the whole regions of Earth. Starting point was related with the competences and experiences within the group activity, contribution our own knowledge and expertise and acted as workshop leaders and lecturers, participated in discussions and presented posters on different themes, updates knowledge and broadened their views. All "sailers" participated actively and they were working with the topics and presentations.

The Sustainable Water Management courses are master's level and focusing on sustainable use of water and water resources in the Baltic Sea region. Students should preferably have a background in e.g., hydrology, geology, physical geography, agronomy, soil science, forestry or environmental engineering (Jankauskas et al. 2004). The full course consists of three modules, which can be studied separately:

The Baltic Waterscape
Water Use & Management
River Basin Management

This course can also be studied as Internet course at Uppsala University.

The Community Development courses are master's level courses focusing on sustainable community development and urban planning (Barinova et al. 2019) with an emphasis on the Baltic Sea region. The courses are interdisciplinary, problem oriented and preparatory for a professional career. The course includes three separately modules:

The City
Building Sustainable Communities

Urban Sustainability Management

The Environmental Management Systems and Certification courses address management in all kinds of organizations, although industries are in focus. They are well adapted for competence development of professionals. The course consists of four modules:

Policy Instruments for Environmental Management
Cleaner Production and Technologies
Product Design & Life Cycle Assessment

Text-book in a series on Environmental Management also was published (Weiß and Bentlage Weiß and Bentlage 2006).

Ecosystem Health and Sustainable Agriculture

This is new educational courses from Baltic University Programme, which was realized in cooperation with Envirovet Baltic programme (2008–2011) to transfer our knowledge on sustainable land use, rural development, and management to teachers, students and professional not only in the Baltic Sea Regions, but also in the Great lake Regions of USA and Canada, because participated in this projects several experts from American universities. The courses address the field of rural development, sustainable agriculture and animal health pertaining to the Baltic Sea Region and to some degree also the Great Lakes region as comparison with USA experience (Jakobsson 2012; Beasley and Adkesson 2012; Karlsson and Ryden 2012).

The marine and coastal terrestrial zones of Earth in different regions its affected by many different toxic pollution from agriculture and waste—water treatment plants, as well as discharge from industries, and waste facilities leads to eutrophication, for instance, in Curonian and Vistula Lagoons, and Baltic Sea too (Wulf et al. 2007; Barinova et al. 2012). In this case was important to increase our knowledge of and improve the management of land and water basins to prevent pollution from agriculture and related activities in rural and urban areas. It is also of great significance health conception to understand and prevent environmental problems for the future.

This example of cooperation it represents a possibility to combined efforts and positive effects of convergence knowledge from agronomists, veterinariens, nature and social geographers, biologist, chemists, economists, business and policy experts (Fedorov and Krasnov 2012; Barinova et al. 2015; Gaeva et al. 2019). It also represents a new way on sustainable agriculture and its parts in the rural ecosystems. There will also be new substantial knowledge on such subjects as land use and rural development, climate change, ecosystem health and the interaction between the wild and domestic animal population, as well as public health components and poverty alleviation (Beasley and Adkesson 2012).

The course package is consist of three modules:

Sustainable Agriculture
Ecology and Animal Health
Rural Development and Land Use

The first are more basic level information, it can be used for students studying at most faculties and could also be attractive for administrators from ministries, government offices or municipalities, advisors and managers. The second and third modules their suggesting for studying more profound agronomy, veterinary, public health, geography, biology, wildlife management or similar subjects. Three books of course package was published in 2012.

In accordance with accepted by participants of these Project the Baltic 21 definition of Sustainable Agriculture contributes significantly to the society of the future. Sustainable agriculture is the production of high quality food and other agricultural products/services in the long run with consideration taken ti economy and social structure, in such a way that the resource base of non-renewable and renewable resources is maintained. Educational materials its available to all universities, in the drainage basin of the Baltic Sea. This project is so for financed by SIDA and the Swedish Institute.

Crisis Management Challenges in Kaliningrad

This Project (2008–2014) was supported by International Association for the Promotion of Cooperation with scientists from the Independent states of the former Soviet Union (INTAS) of years (2008–2014) of cooperation and finished by monograph publication (Krasnov et al. 2014). In this book, five crisis case studies are analysed of events that occurred after the collapse of the Soviet Union.

At 13 May 2015 in Uppsala University in connection with participation in the International Conference on "Teachers Competence Development Training in Education in Sustainable Development: steps to a Sustainable Future" the last certificate was given to me.

This conference included some invited lectures, for instance prominent American futurologist Dennis L. Meadows with new presentation "The Myth of Technology and Sustainable Development", which was very pessimistic about technological solution of all global problems (demographic, poverty, environmental, etc.).

But other presented topics from Baltic countries was more positive and focused to:

• Experimental Education for Sustainable Development
• Teaching for Quality Learning at University
• Alternative teaching methods
• Education Sustainable Development as integral part of teaching.

German-Russian Initiatives

From 2016 was started new German-Russian initiative on joint study "Climate Change and risks for Environmental Security" by experts from Martin-Luther University in Halle-Wittenberg (Germany) and Immanuel Kant Baltic Federal University in Kaliningrad (Russia). The first seminar was focused to regional aspeds of climate change and needs analysis for agriculture security. The problem of alternative sources of energy for local needs also was discussed. Participants of new initiative presents some cases on civil activity instruments for climate defence. By Dr. Dara Gaeva was suggested idea on Internet—platform "Climate Change—Adaptation to these phenomenon in Kaliningrad" organization, which was supported by all participants (Krasnov et al. 2013).

This platform urgently need for local people in the context of very fast and dangerous storms, floods and other negative consequences of modern climate changes, which generate by Atlantic centers of cyclonic dynamics more and more frequently in the last decades. This initiative it open for free discussion, interview, expert estimations, new data and results of study publications etc., all thinks on region and climate generated transformation of soil and waters quality, transport communications situations and so one.

Proposals for Possible Future Cooperation in the South Baltic Region

In accordance with previous experience I would like to suggest for discussion a new typology of South Baltic protected coastal areas development (see Table 1).

To Common Data Base

Hydrometeorological stations, sanitary-hygienic inspectors, the forestry and fishing service and the water utilization boards are responsible for monitoring of various components of the environment in the Baltic Sea region. Observations of air, soil, plants and surface and groundwater quality have regularly been carried out for 20–25 years (Ryden et al. 2003). Furthermore, birds—their numbers and migration patterns—and some mammals are being studied. Shore erosion, as well as other coastal processes, are controlled annually in the Baltic coastal zone. Concentrations of metals, hydrocarbons and other chemical substances, as well as radioactivity, are being measured. Water monitoring was carried out in the by, inter alia, the Lithuanian Hydrometeorological Laboratory in Klaipeda and Russian Atlantic Scientific Research Institute of Marine Fisheries and Oceanography (AtlantNIRO). Altogether, 30 permanent monitoring stations have been in operation in the lagoon areas.

Table 1 Typology of South Baltic protected coastal areas (from Krasnov 1993, with essential remarks)

Name of area	Category	Aim of protection	Country
Terrestrial area Curonian Spit	National park (Russian part)	Conservation of dunes ecosystem, recreation, scientific, research, education	Russia (Kaliningrad) Lithuania
Vistula Spit	Landscape park (Russian part)	Conservation of dunes ecosystem, recreation, scientific, research, education	Russia (Kaliningrad) Poland
Estuarian area Curonian Lagoon	Fish reserve	Aquaculture, fisheries, recreation	Russia (Kaliningrad) Lithuania
Vistula Lagoon	Fish reserve	Aquaculture, fisheries, recreation	Russia (Kaliningrad) Poland
Marine area Curonian slope	Marine park	Exploration Marine park	Russia (Kaliningrad) Lithuania
Vistula slope	Marine park	Exploration Marine park	Russia (Kaliningrad) Poland

An exchange of data between this centre and similar institutions in other countries around the Baltic would be most valuable and, eventually, a common data base for the southeastern Baltic Sea could possibly be established. In the Baltic regions there is, however, an urgent need for up-to-date measuring equipment and sensors for field observations, and equally modern instruments and trained personnel for analyzing data collected at our analytical laboratories. Self-evidently, international intercalibration of methods and practices is also necessary, to develop in accordance with HELCOM declaration (Kropinova 2009; Krasnov and Kropinova 2017).

Conclusion

Despite the long-term efforts of the Baltic University Programme participants environmental education and research of actual human problems in many respects keep ineffective traditions (standard lecturing, preparation of multi-volume manuals, series of seminars and practical classes). The early proposed solutions for the modernization of these processes are often local. A fundamentally different approach is opened by real interdisciplinary "cross-cutting" methods of research and development, which were initiated by Russian scientists V. V. Dokuchaev, V. I. Vernadsky and their western followers (L. Margulis, D. Sagan, J. Lovelock et al.). Students' conscious choice of appropriate training modules and areas of research will help them to achieve the desired results faster, than using previous formal procedures only.

References

Barinova G, Kokhanovskaya M (2015) Climate change monitoring on the basis of phytopheno-logical observations in Kaliningrad region. Pollut Atmosph 226:2268–3798. https://doi.org/10.4267/pollution-atmospherique.5009

Barinova G, Koroleva Yu, Krasnov E (2012) Indicative modeling and spatial evaluation of air pollution risk. In: Kremers H, Susini A (eds) Risk models and applications, Collected Papers. CODATA-Germany, Berlin, pp 23–34

Barinova GM, Krasnov EV, Gaeva DV (2015) Changes of South Baltic Region climate: agroecological challenges and responses. In: Handbook of climate change adaptation. Springer, Berlin, pp 1635–1655. https://doi.org/10.1007/978-3-642-40455-9_17-1

Barinova GM, Ushakova LO, Gaeva DV (2019) Climate change in cities: the environmental effects of climate anomalies. In: Climate change in urban areas: forms and strategies of adaptation, with special consideration of the role of the Russian allotment gardens: proceedings of international conference: scientific electronic edition. Immanuel Kant Baltic Federal University Press, Kaliningrad, pp 62–69

Beasley VP, Adkesson AM (2012) Wildlife and ecosystem health. In: Norrgren L, Levengood JM (eds) Ecosystem health and sustainable agriculture 2: ecology and animal health. Baltic University Press, Uppsala, pp 13–26

Fedorov G, Krasnov E (2012) Agriculture in Kaliningrad. Case study Russia. In: Jakobsson C (ed) Ecosystem health and sustainable agriculture. The Baltic University Press, Uppsala University, pp 448–450

Gaeva DV, Barinova GM, Krasnov EV (2019) Adaptation of Eastern Europe regional agriculture to climate change: risks and management. In: Leal Filho W, Trbic G, Filipovic D (eds) Climate change adaptation in Eastern Europe. Climate change management. Springer, Cham. https://doi.org/10.1007/978-3-030-03383-5_21

Jakobsson C (2012) Definitions of the ecosystems approach and sustainability. In: Jakobsson C (ed) Ecosystem health and sustainable agriculture 1: sustainable agriculture. Baltic University Press, Uppsala, pp 13–15

Jankauskas B, Jankauskiene G, Fullen MA (2004) Erosion-preventive crop rotations and water erosion rates on undulating slopes in Lithuania. Can J Soil Sci 84(2):177–186

Karlsson I, Ryden L (2012) Rural development and land use. In: Rydén L, Karlsson I (eds) Ecosystem health and sustainable agriculture, 3rd edn. Baltic University Press, Uppsala, p 318

Koroleva Y, Revunkov V (2017) Air pollution monitoring in the South-East Baltic using the epiphytic lichen Hypogymnia physodes. Atmosphere 8:119. https://doi.org/10.3390/atmos8070119

Koroleva Y, Napreenko M, Baymuratov R, Schefer R (2019) Bryophytes as a bioindicator for atmospheric deposition in different coastal habitats (a case study in the Russian sector of the Curonian Spit, South-Eastern Baltic). Int J Environ Stud. https://doi.org/10.1080/00207233.2019.1594301

Krasnov E (1993) Present and future nature protection in Kaliningrad. WWF Baltic Bull 6:11–14

Krasnov E (1994) Environmental science and management studies in University. In: Environ-mental science and management studies yearbook University of Latvia Ecological Centre Publisher "Vide", Riga, pp 171–177

Krasnov E, Kropinova E (2017) The combined effects of education and research on sustainable development in the Immanuel Kant Baltic Federal University (Russia, Kaliningrad). In: Leal Filho W et al (eds) Handbook of theory and practice of sustainable development in higher education, vol 2, Springer International Publishing AG 2017, pp 413–423

Krasnov E, Barinova G, Gaeva D (2013) Regional aspects of climate in the southeast Baltic region in connection with global changes. In: Reckermann M, Köppen S (eds) Conference proceedings of the 7th study conference on BALTEX, Borghoim, Island of Öland, Sweden, 10–14 June 2013, pp 73–75

Krasnov E, Karpenko A, Simons G (eds) (2014) Crisis management challenged in Kaliningrad: Monograph Farnham. Ashgate

Kropinova E (2009) Zoning of the Kaliningrad Region for the purpose of tourism and recreation development of trans-border cooperation. Tiltai 3:117–128

Lindroos P (2013) The Baltic University Programme in 2013. Annual Report 2013, Uppsala Centre for Sustainable Development. Uppsala, Sweden, pp 3–9

Ryden L, Migula P, Andersson M (eds) (2003) Environmental science. Baltic University Press, Uppsala

Weiß Ph, Bentlage J (2006) Environmental management systems and certification. Book 4 in a series on Environmental management. The Baltic University Press, Uppsala, p 262

Wulf F, Savchuk OP, Sokolov A, Humborg C, Morth C (2007) Management options and effects on a marine ecosystem: assessing the future of the Baltic. Ambio 336(2–3):243–249

Zaporozhsky DG, Krasnov EV (1996) Cleaner production training at the Kaliningrad universities. In: The proceedings of first cleaner production international conference Rydzyna, Poland, 18–21 Nov 1996, pp 223–227

The Environmental Impact of Shipping in the Baltic Sea Area

Małgorzata Pacuk

Abstract

The aim of the article is to present the main forms of the impact that shipping has on the environment and to discuss legal and organisational solutions and regulations in place to increase the safety of shipping on the Baltic. The methodology applied in the study is induction. Research generalisations were made on the basis of the recent studies results and statistical data. The growing shipping volumes in the Baltic Sea area, especially in crude oil and petroleum products, generate an increased risk of marine incidents which can bring about spills of harmful chemical substances. The shipping route with the highest traffic volume leads north-east to the Gulf of Finland ports. The three largest Baltic ports are located there, with 206 million tons of cargo handled in 2016. Most marine accidents take place in a close distance to the shore. About 75% of the collisions involving tankers end up with oil spills. Maritime transport has a negative influence on the environment also during the daily, failure-free operation of ships. Thanks to the status of a special area of the Baltic, it is possible to introduce stricter regulations on the basin.

Keywords

The Baltic Sea · Maritime traffic · Shipping accidents · Sea pollution · Safety of shipping

M. Pacuk (✉)
Department Regional Development Geography, University of Gdańsk,
Gdańsk, Poland
e-mail: geomp@ug.edu.pl

Introduction

The Baltic Sea is one of the most intensively used water bodies in the world. Baltic ports annually handle ca. 870 million tons of cargo, shipped mainly by bulk carriers and container ships. At the same time, due to the specific features of this sea—mainly shallow depth and a limited exchange of water with the open ocean, low salinity and seasonal ice cover—it is especially sensitive to pollution. This means an increased hazard for the Baltic marine environment both during the daily operation of ships and in failure situations which result in spills of harmful chemical substances.

The article discusses environmental hazards associated with the operation of ships and examples of the latest legal solutions and mechanisms aimed at reducing the detrimental impact of shipping on the condition of the seas and the atmosphere. Since the entry into force of the MARPOL 1973/1978 Convention,[1] the primary act of the international law to regulate the problem of environmental pollution from ships, a significant yet only partial improvement of the situation has been recorded in this regard.

These topics were previously described in the literature on the subject (Fabisiak 2008; Bogalecka 2012; Kupiński et al. 2013; Sormunen et al. 2016; Claremar et al. 2017). This article focuses on the most current matters in this regard.

The Baltic Sea as an Area of Intensive Shipping

Since the mid-nineties of the twentieth century, there has been an enormous increase in the activity of the shipping sector caused by economic growth in the Baltic Sea area and by the development of ports, mainly oil terminals in the Gulf of Finland. The cargo structure has a predominance of bulk (more than 44% of goods shipped on the Baltic are liquid bulk, more than 24%—dry bulk), with the most important cargo in the Baltic being crude oil. According to forecasts, crude oil shipments will increase by more than 60% by 2030 (Matczak 2010).

The Baltic Sea plays an especially important role in Russia's transport policy due to the scale of commercial links between this country and its European partners, carried out through multimodal transport with the dominant share of maritime shipping. The growing foreign trade turnovers between Russia and European countries increase the role of Russian shipping on the Baltic, mainly in exports. The Baltic Sea is an area of focus for the main streams of Russian energy resources addressed to Western markets.

The main Russian ports are located in the Gulf of Finland: Ust-Luga, Primorsk and St. Petersburg; at the same time, these are 3 largest Baltic ports, with 206 million tons of cargo handled in 2016 (180 million tons in 2012). As per the principles of the Russian maritime policy, the Baltic Sea is to become the key transport route for Russian crude oil: by 2030, 40% of all the deliveries of this raw

---

[1]International Convention for the Prevention of Pollution from Ships.

material to EU member states will go through the Baltic (95% through the ports at Primorsk and Ust-Luga) (Pacuk 2015).

In 2013, the conventional border between the North Sea and the Baltic was crossed by 57.6 thousand ships (HELCOM 2014). Each of them is a potential source of pollution to the marine environment and the atmosphere. Ship traffic is the most intensive in the Danish straits but the majority of shipping is not limited to this area. The shipping route with the greatest intensity of traffic leads north-east to the Russian ports in the Gulf of Finland.

Impact of Failure-Free Ship Operation on the Environment

Baltic Sea shipping has been covered by international acts of law and regional projects.[2] Their provisions are there to increase the safety of shipping and at the same time reduce the risk of accidents at sea. The most important international conventions include SOLAS[3] and MARPOL 1973/1978. MARPOL is the primary legal act which regulates the problem of marine environment pollution by harmful substances discharged from ships during their regular daily operation. MARPOL's Annexes specify detailed requirements[4] on matters including: prevention of pollution by oil and oily water, control of pollution by noxious liquid substances in bulk, prevention of pollution by harmful substances carried by sea in packaged form, prevention of pollution by sewage from ships, pollution by garbage discharge from ships, prevention of air pollution from ships. In line with the Convention, any discharge of oily water, chemical residue, garbage or any other pollutant which may be found on a ship should take place in ports, in specially adapted facilities.

Shipping is a significant source of petroleum hydrocarbons in the marine environment. The marine environment is polluted with these compounds not only due to vessel-related disasters but also during regular failure-free operation of ships. Such spills are a significant hazard to the environment but they have a minor importance within the scale of the whole water body. They are most often caused by human error and equipment failure on ships, as well as discharges during ship operation. Recently, there has been an increase in the number of instances of deliberate oil pollution: the largest quantities of illegally discharged pollutants are observed in the Sound (Øresund/Öresund) and in the locations associated with the main shipping routes on the Baltic (Fabisiak 2008).

---

[2]The very poor and still deteriorating condition of the Baltic Sea was identified already in the 1970s as a serious threat to the social and economic growth of the cities located on the Baltic. The joint effort to protect the waters and living resources of the Baltic by all the countries of the region began with the signing of the Gdańsk Convention (1973) and the Helsinki Convention (1974). The latter resulted in establishing the Baltic Marine Environment Protection Commission—Helsinki Commission (HELCOM).
[3]International Convention for the Safety of Life at Sea, 1974.
[4]Special, stricter rules are in force in special areas. As an especially endangered water body due to its environmental sensitivity and intensity of ship traffic, the Baltic Sea has received the status of such an area.

Accidents and Incidents at Sea

There has never been a catastrophic spill in the Baltic Sea, with a scale comparable to the largest events of this type, although in the past there were accidents which resulted in larger spills. Since the 1960s, more than 40 serious oil spills have been recorded on the Baltic, as a result of which more than 100 tons of petroleum compounds were released into the sea in one go (Fabisiak 2008). Even though the risk of oil spills increases with a growth in the shipping of crude oil and petroleum products. The number and estimated volume of oil spills detected in the Baltic has been constantly decreasing. The total number of mineral oil discharges observed during aerial surveillance decreased from 472 to 53 in the years 2000–2016. The result of 2016 is the lowest ever recorded in the area. 89% of spills detected in 2016 were smaller than 1 m^3 and the largest oil spill was estimated to be 1.6 m^3. This is the result of constant intensive aerial surveillance of ships on the sea (HELCOM 2017).

In 2013, 150 sea accidents and incidents were recorded on the Baltic, which is the highest figure in a 10 year period and an increase of 15% compared to 2010. Most of them took place within a close distance of the shore: in ports of on port approach (45%), with 34% recorded at open sea.[5] The most frequent causes of accidents recorded in 2013 include: collisions (39%) and groundings (29%). In 2004–2013, the percentage of collisions stayed at a similar level, with a decrease recorded in the percentage share of groundings. Within the studied period, collisions were the most frequent in the Danish straits, the Gulf of Finland and the Gulf of Bothnia. The types of ships participating in accidents are predominantly cargo vessels (49% in 2013), with the share of tankers close to 10% in 2013. According to the 2004–2013 figures, 4.7% of recorded sea accidents included a leak of harmful substances to the sea (heavy fuel oil (mazout), diesel fuel, hydraulic oil) (HELCOM 2014).

Due to the specific nature of the Baltic Sea (a closed sea, long water exchange time), chemical substance leaks generate consequences which are much more noxious to the environment than in the case of open water bodies.

Legal Regulations and Technical Measures to Combat Environmental Pollution

By the decision of the International Maritime Organisation (IMO), in 2005 the Baltic Sea received the status of a Particularly Sensitive Sea Area (PSSA) which is covered by special protective standards to enable an increased control over movements of vessels. The most important ones are: areas to be avoided, a traffic separation scheme, or a designation of independent routes for ships which travel north and south, and a pilotage requirement in the Danish straits.

As one of the world's most important water bodies for ferries and Ro-Ro ships, the Baltic Sea has been designated a Sulphur Emission Control Area (SECA). The Sulphur Directive, in force here since 1 January 2015, makes it mandatory for ship

---

[5]There are no detailed figures on the remaining 21% of hazardous shipping events.

owners to use low sulphur fuels (reduction of the previous limit of 1–0.1%). In order to meet the new cost-intensive requirements,[6] some vessels are being upgraded to enable exhaust gas desulphurisation. Some ship owners invest in new LNG-powered ferries to allow them to completely eliminate sulphur emissions.

Due to the constant threat to the environment associated with intensive shipping, Baltic Sea countries take measures to reduce and combat sea pollution, which they are obliged to do by international conventions. The tasks of Search and Rescue (SAR) services[7] include, for example, to coordinate missions to remove threats and pollutants from the marine environment, to remove spills of harmful chemical substances formed as a result of maritime accidents and disasters, to look for and bring up harmful substances and cargo. Specialist units dedicated to combating oil spills at sea are based in the main Baltic Sea ports and in the ports located in the direct vicinity of its main shipping route. However, no Baltic Sea country alone has sufficient resources to take appropriate measures if there is a catastrophic spill on the Baltic. This is why every year (starting from 1990), there are international Balex Delta exercises attended by specialist units from the Baltic Sea countries. The measures to protect the marine environment of the Baltic are also joined by patrol aircraft which are part of the border guard, navy or air force aviation units. The efficiency of such measures is proven by the fact that, since 2000, a continuous drop has been recorded in observed spills. The vast majority of these spills did not constitute a serious enough hazard to make it necessary to take combating action (Tarkowski 2016).

Summary

The Baltic Sea is a water body with the greatest shipping intensity in the world. A high and growing share in Baltic shipping is taken up by Russian crude oil and petroleum products. This means that there is an increased risk of marine accidents resulting in spills of harmful chemical substances. The shipping of petroleum derivatives is the greatest threat to the Baltic because most of collisions with tankers end up in an oil spill. It follows from observation that most frequently these spills are very small or small and are partially biodegraded. In the case of a catastrophic spill on the Baltic, international collaboration is necessary, along with the joint use of the resources of each state to take measures corresponding to the scale of the hazard.

The necessity to regulate maritime activities in a very detailed way by means of administrative rules, technical and quality standards aimed at protecting the marine environment, becomes especially important in a water body as unique as the Baltic

---

[6]Fuel is the most expensive component of vessel operation costs. As the price of low sulphur fuel is 30–50% higher than the price of traditional fuel, it is estimated that transport costs will increase by EUR 2–9 per ton of cargo (Christowa et al. 2014).

[7]In Poland, the obligations under the International Convention on Maritime Search and Rescue (SAR) are performed by the Maritime Search and Rescue Service (MSPiR).

Sea. The status of a special area awarded to the Baltic makes it possible to introduce stricter regulations to reduce pollution loads introduced into the sea from vessels. More than 10 years from the entry into force of MARPOL Annex VI, which regulates the emission of specific pollutants from vessels to the atmosphere, it is now assessed that air conservation has become as important an aspect of international activities as the protection of the marine environment. In the case of a water body with a high intensity of shipping, such as the Baltic, exhaust gas emissions are greatly cumulated. The introduction of the Sulphur Directive, aimed at reducing atmospheric pollution from vessel operation, may have an influence including an accelerated development of new technologies in the production of marine fuels.

References

Bogalecka M (2012) Bezpieczeństwo transportu morskiego w regionie Morza Bałtyckiego. Zarządzanie i Finanse 3:572–580

Christowa Cz, Christowa-Dobrowolska M (2014) Wpływ dyrektywy siarkowej Parlamentu Europejskiego na konkurencyjność przedsiębiorstw żeglugi promowej na Bałtyku. Koncepcja badań naukowych. Logistyka 3:1025–1036

Claremar B, Haglund K, Rutgersson A (2017) Ship emissions and the use of current air cleaning technology: contributions to air pollution and acidification in the Baltic Sea. Earth Syst Dyn 8:901–919. https://doi.org/10.5194/esd-8-901-2017

Fabisiak J (2008) Zagrożenia ekologiczne Bałtyku związane z zanieczyszczeniami chemicznymi —węglowodory. Zeszyty Naukowe Akademii Marynarki Wojennej 3:7–28

HELCOM (2014) Annual report on shipping accidents in the Baltic Sea 2013. http://www.helcom. fi/Lists/Publications/Annual%20report%20on%20shipping%20accidents%20in%20the%20Baltic %20Sea%20area%20during%202013.pdf. Accessed 30 June 2018

HELCOM (2017) Annual report on discharges observed during aerial surveillance in the Baltic Sea 2016. http://www.helcom.fi/Lists/Publications/Annual%20report%20on%20discharges% 20observed%20during%20aerial%20surveillance%20in%20the%20Baltic%20Sea%202016.pdf. Accessed 30 June 2018

Kupiński J, Michalak J, Fabisiak J (2013) Ryzyko rozlewów produktów ropopochodnych i innych substancji niebezpiecznych w akwenie Morza Bałtyckiego. http://www.logistyka.net.pl/ bank-wiedzy/transport-i-spedycja/item/download/75923_fe3ab559ee407c52e4f6fed50b8a7765. Accessed 25 Oct 2017

Matczak M (2010) Gospodarka, handel i transport bałtycki w pierwszej dekadzie XXI wieku. In: Europa Bałtycka. Przeszłość, teraźniejszość, nowe wyzwania. University of Szczecin, Szczecin, pp 47–56

Pacuk M (2015) Porty rosyjskie Zatoki Fińskiej. Logistyka 3:3685–3691

Sormunen O-V, Hänninen M, Kujala P (2016) Marine traffic, accidents and underreporting in the Baltic Sea. Sci J Marit Univ Szczecin. https://doi.org/10.17402/134

Tarkowski M (2016) Zwalczanie zagrożeń i zanieczyszczeń środowiska na polskich obszarach morskich. Problemy Transportu i Logistyki 4:105–114

The Reconstruction of the Formation of Lakes and Wetlands and the Related Sedimentation Processes in the Russian Segment of the Vištytis Upland

Yuri A. Kublitsky◉ and Dmitry Subetto◉

Abstract

The Vištytis Upland is a terminal moraine ridge featuring kame hills. There are numerous lakes and wetlands in its depressions. We collected the samples of bottom sediments from a number of lakes and wetlands located at different altitudes. Using the results of the lithological analysis and radiocarbon dating of the lakes' bottom sediments and those of a loss-on-ignition test, we identified the main stages of sedimentation. We reconstructed the formation and development of the lakes of the Vištytis Upland from the moment of deglaciation to the present. Two groups of lakes were distinguished by genesis: (a) the lakes formed in the inter-moraine or inter-kame depressions flooded by melt water; (b) the lakes formed in the thermokarst processes. The lakes of the first type are located at altitudes over 170 m and date back to the Younger Dryas, whereas the lakes of the second type are located below 170 m and their formation occurred in the Younger Dryas—the Preboreal. We conclude that the wetlands located west of Lake Vistytis were the bays of the lake in the Late Pleistocene.

Y. A. Kublitsky (✉) · D. Subetto
Department of Geography, The Herzen State Pedagogical University of Russia,
St. Petersburg, Russia
e-mail: Uriy_87@mail.ru

D. Subetto
e-mail: subetto@mail.ru

D. Subetto
Institute for Water and Environmental Problems, Siberian Branch of the Russian Academy
of Sciences, Barnaul, Russia

D. Subetto
Immanuel Kant Baltic Federal University, Kaliningrad, Russia

G. Fedorov et al. (eds.), *Baltic Region—The Region of Cooperation*,
Springer Proceedings in Earth and Environmental Sciences,
https://doi.org/10.1007/978-3-030-14519-4_14

Keywords
Bottom sediments of lakes · Glacial lakes · Lake formation · Vištytis Upland ·
Kaliningrad region · Sedimentation · Late glacial · Holocene

Introduction

Studying the environmental transformations of the past is highly relevant, since
only our knowledge of the past changes can shed light on how geosystems will be
transformed in the future. Lakes are unique natural objects, the bottom sediments of
which keep records of the past natural and climatic conditions, ranging from local to
global in terms of space and from annual to centennial and to millennial in terms of
time. The data on the lithologic structure of lakes' bottom sediments makes it
possible to perform palaeogeographic reconstructions and to study the sediment
formation and the changes in the hydrological regimes that water bodies went
through in the past (Subetto 2009). A poorly explored area, the Vištytis
hilly-moraine ridge occupies a special place in regional palaeogeographic studies.
For the reconstruction of the sedimentation processes, we chose several lakes and a
bog located at different altitudes. In this article, we present a lithological description
of Lake Chistoe and the peat-bog of Shombrukh, a digital geological model of the
Vištytis Upland, and a reconstruction of the palaeobasins formation.

The Study Area

The studied lakes and bog are located the Vištytis Upland (Fig. 1), which formed
during the Baltija (Pomeranian) stage of the Last (Weichselian) Glaciation (Guobytė
and Satkunas 2011; Raukas et al. 2010). The Vištytis Upland was one of the first
areas of today's Kaliningrad region to be freed from the glacier (Kublitskii 2016).
Thus, it contains traces of the oldest lake sediments. Although the topography of the
Vištytis Upland is mostly of an inherited nature, there are some inversions. The
altitudes are about 50 m in the northwest and 200–300 m in the southeast (Orlenok
et al. 2001). In terms of geomorphology, the territory is characterised by facies of
push—and thrust-block-types with the prevalence of water-glacial forms, such as
moraines, kames, eskers, sandurs, and lacustrine-glacial plains. There is a palustrine
wetland comprised by flat and slightly arching upland palustrine plains. With their
feet coalescing, the hills form ridges divided by depressions. These depressions are
mostly swamped and, in some, lakes are formed.

The climate of the Vištytis Upland is more continental than in the rest of the
Kaliningrad Region. The average annual quantity of precipitations is 700 mm, the
average annual temperature is +6.5 °C, the average January temperature is −4.5 °C,
and the average July temperature is +17.5 °C (Litvin 1999). The predominant soil

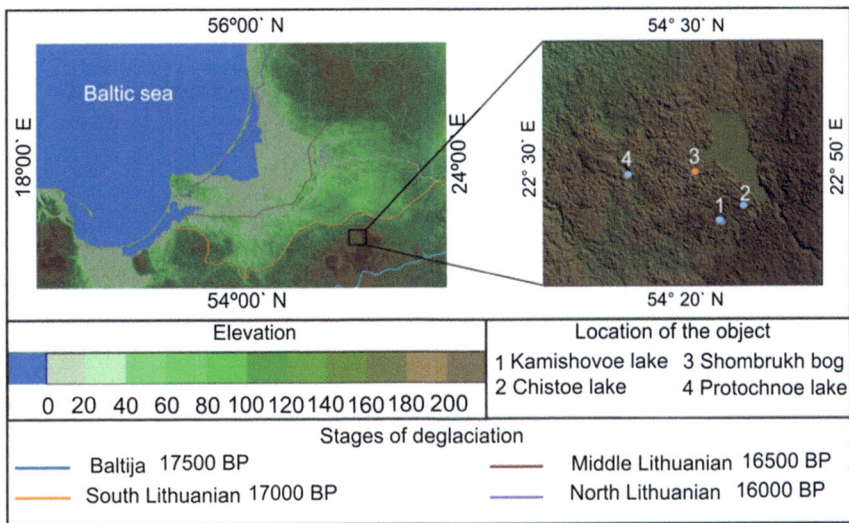

Fig. 1 The location of the study areas, and the stages of the last glaciation

types are brown and cryptopodzolic. Besides, in the depressions, there is marsh soil formed by excessive precipitations. The vegetation cover is diverse: pines and spruces comprise over half of the forest, whereas a slightly smaller area is occupied by small-leaved trees. Broad-leaved forests account for about 10% of the territory (Orlenok et al. 2001).

Methods

Field research was conducted from 2011 to 2015. The bottom sediment cores were sampled by peat corer (the sampler diameter was 7.5 cm for the upper horizons and 5 cm for the denser lower horizons and bogs). A primary visual lithological stratigraphic description was prepared on site, after which each meter column of sediments was placed in plastic containers for transportation. A detailed description, sampling, radiocarbon dating, and a loss-on-ignition (LOI) test were carried out in the laboratory, employing standard techniques (Santisteban et al. 2004). The collected samples were analysed using the conventional advanced radiocarbon method at the Laboratory of Palaeography and Geochronology of the Quaternary period (Saint-Petersburg State University, Russia), at the A. E. Fersman Laboratory of Environmental Geochemistry (The Herzen State Pedagogical University of Russia), and at a laboratory in Poznan (Poland). To determine the type of the bottom sediments, a soil organic matter classification was used: a LOI value of below 2% is characteristic of clay, 2–6% of gyttja clay, 6–20% of clayish gyttja, and 20% of gyttja proper (Miettinen et al. 2007). A digital terrain model was created using the Global Mapper 16 and the Golden Software Surfer 13 applications.

Results

Lithological stratigraphy A visual lithological description and the chronology of the bottom sediments of Lakes Kamyshovoe, Protochnoe, Chistoe, and the peat-bog of Shombrukh were published earlier (Kublitskii et al. 2014a, b, 2016).

Geochronology A detailed chronological dating was carried out only for Lake Kamyshovoe (Kublitskii et al. 2014a). For Lake Protochnoe, there is one radio-carbon date from the lower horizon that makes it possible to pinpoint the beginning of the lake sediment formation. Using the dating results, the average rates of sedimentation were calculated (Table 1).

Discussion

These water bodies are located at different altitudes in close proximity to each other. The greatest distance between the objects (Lake Protochnoe and Lake Chistoe) does not exceed 8 km. Nevertheless, the lower parts of the bottom sediments sections differ, which indicates that they were not formed under similar conditions (Fig. 2).

Lakes Kamyshovoe and Chistoe, which are located the altitudes of 192 and 207 m respectively, have a similar structure. The lower part of the bottom sediments cores is grey gyttja clay, with a dark-brown gyttja interbed at depths of 1060–1064 cm (Lake Kamyshovoe) and 689–704 cm (Lake Chistoe). This gyttja interbed dates to 13,714 cal. years BP and its formation is associated with the Allerød oscillation (Druzhinina et al. 2015; Kublitskii 2016). A similar horizon was found in the bottom sediments of the lakes in Poland, Lithuania and Belarus (Wachnik 2009; Kabailienė 2006; Novik 2010). The territory under study was effectively ice-free around 16.5–17 cal. k year BP (Hughes et al. 2016), which suggests that sedimentation began in the same period. Therefore, Lakes Kamyshovoe and Chistoe were formed in the depressions of the moraine ridges after the glacier had retreated. From the time of their formation and until the onset of the Allerød, the allochthonous sedimentation type prevailed in the lakes. This sedimentation type was characterised by the predominance of mineragenic sediments over organogenic in the conditions of a cold climate and periglacial vegetation. In the Allerød the bottom sediments structure was transformed—the gyttja clay turned into gyttja, due to the increased role of organic matter in the formation of bottom sediments. The formation of sediments with a high organic content is associated with the improvements in the climatic conditions in the catchment area of the lake.

Table 1 Summary data on the water bodies

The water body and sampling site coordinates	Altitude (m)	The thickness of sediments (cm)	Horizon dated (cm)	Lab. No.	^{14}C dating BP	Cal. age BP	The average sedimentation rate, mm/year
Lake Protochnoe 52°24.122′ N 22°36.458′ E	153	864	831–841	SPb_1223	10,300 ± 100	12,101	0.71
Lake Kamyshovoe 54°22.605′N 22°42.790′ E	192	975	838–839	POZ-60,940	11,600 ± 60	13,417	0.62
Lake Chistoe 54°23.302′ N 22°43.706′ E	207	372	340–344	SPb_1797	10,450 ± 80	12,350	0.3
Shombrukh bog 54°24.207′ N 22°40.622′ E	187	740	720–730	LU-7260	8620 ± 210	9774	0.71

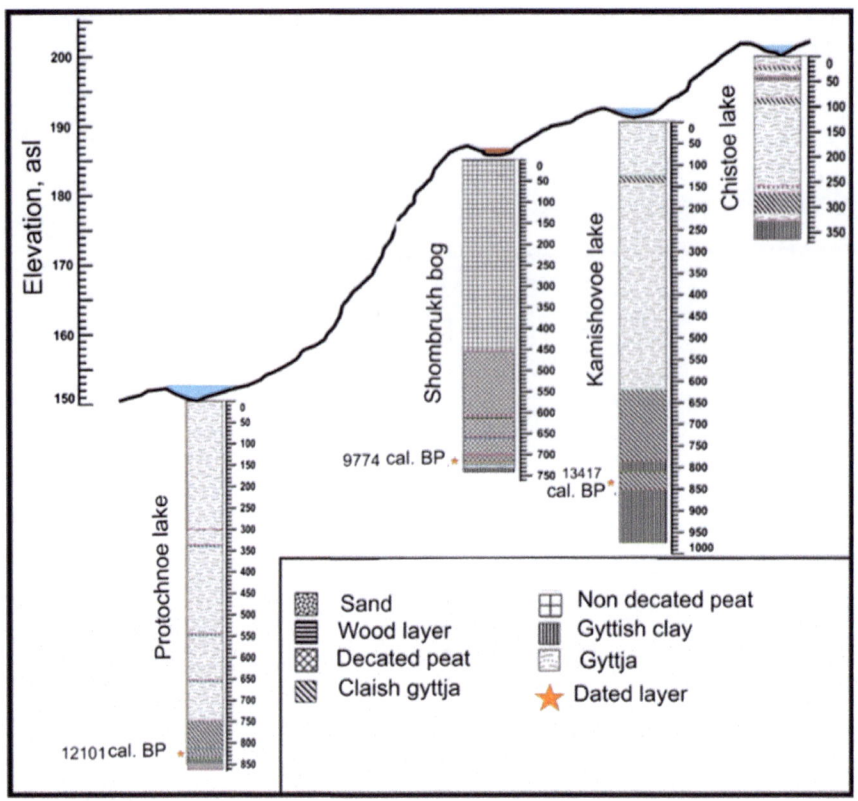

Fig. 2 Lithology, chronology and sedimentation rates of the bottom sediments of the lakes in the Vištytis Upland

Around 12,700 cal. years BP (Druzhinina et al. 2015), the predominantly organogenic gyttja was replaced by predominantly mineral gyttja. This testifies to the growing role of the allochthonous substances in the sedimentation process that is associated with the declining temperatures of the Younger Dryas. At the beginning of the Holocene (11,700–11,000 cal. years BP), the proportion of organic matter in the bottom sediments started to increase. From the Boreal, the lithology of the sediments was represented predominantly by gyttja.

The rates of sedimentation in Lake Kamyshovoe are 2.5–3 times higher those in Lake Chistoe. This is explained by as follows: (1) the samples were taken near the shore, whereas sediments are usually the thickest at maximum depths, were we could not perform sampling for technical reasons; (2) the catchment area of a water reservoir is smaller at greater altitudes, which leads to the reduction in the deposits brought into the basin.

The peat-bog of Schombrukh is located 2 km west of Lake Vistytis, at an altitude of 180 m. The lowermost part of the bottom sediment section from the bog is clay, which indicates that this basin was a part of a glaciolacustrine basin in the

Fig. 3 A reconstruction of the formation of the selected lakes in the Vištytis Upland. **a** 14000–16,000 cal. years BP; **b** 12,700–14,000 cal. years BP; **c** 11,700–12,200; **d** 9774 cal. years BP—present time

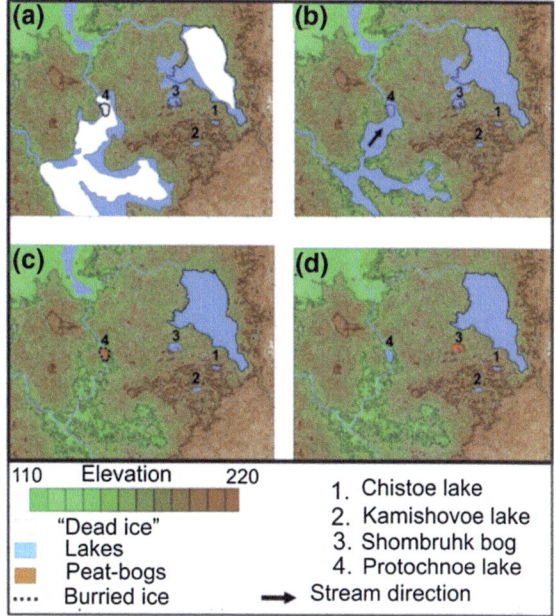

Late Glacial. During the formation of clays, the bog was 7–8 m below the present level (if the thickness of the Holocene peat is subtracted), thus, the bog basin could be flooded with the waters from palaeo-Lake Vistytis. Today, the altitude of the lake is 172 m (Orlenok et al. 2001) but, in the Late Glacial period, it could be 180 m (Fig. 3b). The water level of Lake Vistytis could go down during the drainage of the Baltic Ice Lake, 11,690 cal. years BP (Jakobsson et al. 2007; Björck 1995). Later, in an already isolated small lake, the gyttja and peat started to form (Fig. 3c). The sand layers occurring in the gyttja and even in the lower peat horizons are associated with the drainage and isolation of the bog basin from Lake Vistytis. Above the sand interbed lies a 7-m peat stratum, which reaches the day surface. The stratum formed after 9500 cal. years BP.

Lake Protochnoe is located at an altitude of 153 m. The sediments of the lower part of the section are peat overlain by sand and gravel, the dimensions of which increase with depth (Kublitskii et al. 2014b). This stratigraphy suggests two scenarios of the lake formation. Firstly, the reservoir might have formed by the melting of dead ice blocks. This type of lake genesis is typical for the region under study (Nalepka and Jurochnik 2013). The second scenario is associated with the melting of the buried "dead ice" blocks during deglaciation. To test the second scenario, we created a digital relief model, the analysis of which suggests that the depression located south of the lake could indeed have been covered by dead ice (Fig. 3a). The meltwater streams flowed northward through the lake basin (Fig. 3b). Lake Protochnoe was formed within the area of dead ice melting in the valley of the palaeo-water discharge. The sand and gravel, which were uncovered below the peat, are alluvial sediments. After the dead ice had melted, the swamping of the lake basin began (Fig. 3c). This happened 12,100 cal. years BP, i.e. at the end of the Younger Dryas. The insufficient number of dates precluded an estimate of the precise age of the wetland ecosystem. Since we know that the thickness of the peat is 15 cm, it is possible to estimate the time of the formation of the bog, which is about 50–100 years. As the buried ice melted, the bottom of the basin subsided. As a result, the wetland ecosystem transformed into a lake ecosystem and the process of accumulation of peat was replaced by the that of silt.

Conclusion

All studied lakes and wetlands of the Vištytis Upland formed during the Late Glacial period. By their genesis, they can be divided into two types: (1) the glacial lakes (formed in the inter-moraine or inter-kame depressions by meltwater flooding; (2) the thermokarst lakes. The lakes of the first type are located at higher altitudes. In the lakes of this type, the process of sediment formation began as the glacier started to retreat. In the sediments of the Younger Dryas, the mineral fraction (clay and gyttja clay) predominates. A distinctive feature of the most ancient lakes is that the clay layer contains a gyttja interbed, which is dated to the Allerød interstadial. The reservoirs of the second type are at altitudes below 170 m. The onset of lake sediment formations is associated with the Pleistocene/Holocene transition. The sediments of a part of the uncovered column are peat overlying sand. The area of the peat-bog of Schombrukh might have been flooded by palaeo-Lake Vistytis. Today, the palaeolake is located at an altitude of 172 m. During the Late Glacial, it could have been at 180 m.

Acknowledgements This study was carried out with the support of the Russian Foundation for Basic Research, projects No. 18-05-80087 and No. 18-55-00008.

References

Björck S (1995) A review of the history of the Baltic Sea, 13.0–8.0 ka. Quat Int 27:19–40

Druzhinina O, Subetto D, Stančikaitė M, Vaikutienė G, Kublitsky J, Arslanov Kh (2015) Sediment record from the Kamyshovoe Lake: history of vegetation during late Pleistocene and early Holocene (Kaliningrad District, Russia). Vilnius Balt 28(2):121–134

Guobytė R, Satkūnas J (2011) Pleistocene glaciations in Lithuania. In: Ehlers J, Gibbard PL, Hughes PD (eds) Quaternary glaciations extent and chronology—a closer look. Elsevier, Amsterdam, pp 231–246. http://doi.org/10.1016/B978-0-444-53447-7.00019-2

Hughes A, Gyllencreutz R, Lohne Ø, Mangerud J, Svendsen J (2016) The last Eurasian ice sheets —a chronological database and time-slice reconstruction, DATED-1. Boreas 45:1–45

Jakobsson M, Björck S, Alm G, Andrén T, Lindeberg G, Svensson N (2007) Reconstructing the Younger Dryas ice dammed lake in the Baltic Basin: Bathymetry, area and volume. Glob Planet Chang 57:355–370

Kabailienė M (2006) Late glacial and holocene stratigraphy of Lithuania based on pollen and diatom data. Geologija 54:42–48

Kublitskii YA (2016) Dinamika prirodnykh uslovii yugo-vostochnoi chasti Baltiiskogo regiona v pozdnem neopleistotsene i v golotsene (Dynamics of the natural conditions of the south-eastern part of the Baltic region in the late Neperestocene and in the Holocene). Dissertation, St. Petersburg (in Russ.)

Kublitskii YA, Subetto DA, Druzhinina OA, Arslanov KhA, Skhodnov IN (2014a) Palaeoklimaticheskaya rekonstruktsiya izmenenii prirodno-klimaticheskikh obstanovok v kontse pleistotsena i golotsene v Yugo-Vostochnoi chasti Baltiiskogo regiona po dannym litologicheskogo analiza i dinamike poter' pri proka-livanii (Palaeoclimatic reconstruction of changes in the natural climatic conditions at the end of the Pleistocene and Holocene in the South-Eastern part of the Baltic region according to the lithological analysis and the dynamics of losses during the—living). Obshchestvo Sreda Razvit 2(31):179–184 (in Russ.)

Kublitskii YA, Subetto DA, Druzhinina OA, Masterova NN, Skhodnov IN (2014b) Rekonstruktsiya prirod-no-klimaticheskikh izmenenii v Yugo-Vostochnoi chasti Baltiiskogo regiona v golotsene po rezul'tatam is-sledovaniya donnykh otlozhenii ozera Protochnogo (Kaliningradskaya oblast') (Reconstruction of natural and climatic changes in the South-Eastern part of the Baltic region in the Holocene according to the results of the investigation of the bottom sediments of Lake Protochny (Kaliningrad Region). Elektron Zh Open Bull (Open Sci Bull) 2:1–6 (in Russ.)

Litvin V (1999) Prirodnye landshafty Kaliningradskoi oblasti (Natural landscapes of the Kaliningrad District). Kaliningradskaya Oblast: Prirodnye Resursy, Kaliningrad, pp 141–151 (in Russ.)

Miettinen A, Savelieva L, Subetto D, Dzhinoridze R, Arslanov Kh, Hyvarinen H (2007) Palaeoenvironment of the Karelian Isthmus, the easternmost part of the Gulf of Finland, during the Litorina Sea stage of the Baltic Sea history. Boreas 36(4):441–458

Nalepka D, Jurochnik A (2013) Late glacial and holocene plant cover in węgliny, lubsza plain, south-west Poland, based on pollen analysis. Acta Palaeobot 53(2):191–233

Novik A, Punning JM, Zernitskaya V (2010) The development of Belarusian lakes during the Late Glacial and Holocene. Est J Earth Sci 59:63–79. https://doi.org/10.3176/earth.2010.1.05

Orlenok VV, Barinova GM, Kucheryavyi PP, Ul'yashev GL (2001) Vištytiskoe ozero: priroda, isto-riya, ekologiya (Vištytiskoe lake: nature, history, ecology). Izd. KSU, Kaliningrad (in Russ.)

Raukas A, Wojciech TJ, Stankowski et al (2010) Chronology of the last deglaciation in the Southeastern Baltic region on the basis of recent OSL dates. Geochronometria 36:47–54

Santisteban JI, Mediavilla R, Lypez-Pamo E, Dabrio CJ, Zapata MBR, Garcia MJG, Castano S, Martinez-Alfaro PE (2004) Loss on ignition: a qualitative or quantitative method for organic matter and carbonate mineral content in sediments. J Palaeolimnol 32:287–299

Subetto DA (2009) Donnye otlozheniya ozer: palaeolimnologicheskie rekonstruktsii (Bottom sediments of lakes: palaeolimnological reconstructions). St. Petersburg (in Russ.)

Wachnik A (2009) Vegetation development in the Lake Miłkowskie area, north-eastern Poland, from the Plenivistulian to the late Holocene. Acta Palaeobot 49:287–335

On the Issue of the ILC
in the South-Western Baltic Sea Region

Victor P. Dedkov⬤

Abstract

The Russian Federation has a unique landscape diversity that requires detailed study, rational use and protection. The solution of these problems is impossible both without research, and without educating and involving general public into discussions and decision-making at communal, municipal and regional levels. Unfortunately, there is no unanimity on topical issues of conservation and landscape management. All this hampers the adoption of well-balanced management decisions for the preservation, development and improvement of landscapes both at the level of government structures and at the municipal level. Given this the goal of the article is to attempt at providing a comprehensive analysis of the issue related not only to the solution of problems of landscape protection and rational nature management, but also to possible ways of organizing international landscape centers for conservation of natural, natural historical and cultural landscapes both at the national and at the regional level. The article focuses on the Kaliningrad region as an example which justifies a modular approach to solving such issues, on new technological grounds (MOOC, LMS, EdTech, GreenField). The basic principles of the work of international landscape centers are formulated: scientific, interdisciplinary, mass, open, communicative, institutional, participatory, resourceful.

Keywords

Landscape · Environment · Landscape planning · Development objectives · GIS

V. P. Dedkov (✉)
Institute of Living Systems, Immanuel Kant Baltic Federal University, Kaliningrad, Russia
e-mail: VDedkov@kantiana.ru

© Springer Nature Switzerland AG 2020   137
G. Fedorov et al. (eds.), *Baltic Region—The Region of Cooperation*,
Springer Proceedings in Earth and Environmental Sciences,
https://doi.org/10.1007/978-3-030-14519-4_15

Introduction: Some Prerequisites to ILCs

In the general EU strategy on the Russian Federation and the Russian strategy towards the EU (Moscow 2000), the Kaliningrad Region is a kind of European facade of Russia in the southeast of the Baltic Sea region and is viewed as a "pilot" Russian region in cooperation with EU in the XXI century in various fields: political, socio-economic, scientific, cultural, environmental (Dedkov and Fedorov 2006; Bundesamt für Naturschutz mit schwerpunkt landschaftsplaning 2006).

The area of the Kaliningrad region is slightly larger than 15 thousand square kilometers, the population is about 1 million people, so the population density is among the highest in the country. There are 62 people per one square kilometer, and in the coastal zone it is up to 200 people (Litvin et al. 1999).

Due to the high degree of development of lands that could be classified as natural landscapes, there are no more than 10% of such, about the same amount can be named poorly transformed landscapes. For example, in Russia this figure is more than twice as high. On 80% of the territory, natural ecosystems are destroyed or severely transformed through economic activities (Dedkov et al. 1999, 2004).

Despite this, the land is being seized for various development projects. It is more likely to happen where natural landscapes and ecosystems remain preserved, i.e., in the area of the unique Curonian and Baltic spits, along the coast of the Baltic Sea, the Curonian and Kaliningrad bays, the Vyshtynets Upland. Despite this, there is still something in the Kaliningrad region that requires protection. This is the Curonian and Baltic spit, marsh complexes (Pravdinsky, Gromovsky, Koziy, Bolshaya Mokhovoye), small rivers (Prokhladnaya, Vitushka, Kornevka, Krasnaya), lake complexes (Lake Vishtynetskoe), forests with predominance of coniferous and broad-leaved trees on unique brown forest soils (forest massifs Krasny, Pravdinsky, Ozersky, Gromovsky, Dalny, Zimovniki), a variety of biotopes with a number of rare and endangered species of plants and animals, transboundary territories (the Neman river delta, the Red Forest in the Vishtynetsky region yshennosti), coastal wetlands arrays and some others (Litvin et al. 1999; Dedkov 2004).

Key Issues for Environmental Protection in the Exclave Territory

The uniqueness of the geopolitical position of the Kaliningrad Oblast obliges us to comply not only with Russian environmental legislation, but also with international environmental treaties (Dedkov and Fedorov 2006).

Analysis of socio-economic development and environmental policy makes it possible to say that the existing territorial-planning schemes of nature management do not contribute to the creation of environmentally oriented economic activity, which leads to conflict situations between business entities and nature protection structures both within the Kaliningrad region and outside the exclave. In the end, this disrupts the environment-forming functions, destabilizes landscapes and deteriorates the quality of the parameters of the natural environment. The total area of protected landscape is insufficient to maintain natural balance, biotopic and biological diversity (Ministry of Natural Resources and Ecology of the Kaliningrad Region 2017). The ongoing economic activity jeopardizes the functioning of the already established specially planned protected areas. Local residents are poorly informed and practically do not participate in decision-making on key issues of territorial planning and development.

Promising Solutions to Territorial Environmental Problems

These methodological deficiencies can be minimized and eliminated through the ILC which can be created in Russian federal universities and emphasize the use of landscape planning tools. In the European Landscape Convention, the term is defined as forward planning, which aims to improve, restore and form landscapes (Antipov and Drozdov 2002). Landscape planning in the EU countries is a tool for planning, conservation and care of the landscape, focused on preventive measures; its object is the entire territory where landscape is protected, taken care of and developed (Bundesamt für Naturschutz mit schwerpunkt landschaftsplaning 2006).

Landscape planning is a communicative process involving general public into the decision-making process, thereby strengthening the positions of civil society. Given a qualitatively new background, it is possible to develop a legal framework for environmentally oriented land use and environmental protection; get a large array of versatile knowledge about nature and natural processes in digital format using geo-information technologies (Bundesamt für Naturschutz mit schwerpunkt landschaftsplaning 2006).

Research Capacities and the Awareness Level

Implementing the project on establishing an ILC in the southeast of the Baltic Sea region, in the Kaliningrad Region, is supported by a wide range of information on the state of the natural environment and socio-economic development of the region (Litvin et al. 1999; Dedkov 2004; Dedkov and Grishanov 2010; Ministry of Natural Resources and Ecology of the Kaliningrad Region 2017).

Educational, research and practical institutions of the region employ a good number of highly qualified specialists (biologists, ecologists, geographers, lawyers) capable of solving the tasks of the center.

The mission of the ILC "Kantiana" lies in optimization of ways of co-evolution of man, nature and society as the prerequisite for sustainable development of the Russian exclave in the Baltic Sea region, through education, landscape management culture, education, science and humanitarization of natural scientific knowledge; implementation of the key ideas of academicians V. I. Vernadsky and

N. N. Moiseev about the role of mankind in the biosphere and the ways of their co-evolution.

There are some tasks to implement:

to accumulate full information on territorial (urban) and landscape planning as a methodology background to provide for advisory assistance with project documentation focused on reconciling economic, social and environmental issues, optimizing local green economy and population settlement patterns with a view of nature conservation;

to set up an educational platform in the region through institutes of higher professional education, additional vocational education, advanced training and retraining and develop cultural competencies among the general public;

to use the ILC's potential to strengthen the position of civil society through the establishment of partnership between the region's population, local authorities and business community to implement the sustainable development knowledge platform Rio + 20 "The future we want."

ILC basic principles bear interdisciplinary, mass character; openness; communication; participation; resource; they consider institutional and scientific development.

ILC will follow a module pattern.

Module 1. Education for the exclave sustainable development

1.1. Development of interdisciplinary network educational start-ups in "Landscape Planning", "Sustainable Development" for federal universities of the Russian Federation in some advanced technological platforms (MOOC, LMS, EdTech, GreenField) within the higher professional education;

1.2. Development of interdisciplinary educational start-ups in "Shape and texture in landscape architecture", "Natural and anthropogenic landscapes: urban and recreational", "Natural and anthropogenic landscapes: industrial and transport, geotechnical systems, geo-ecological background for landscape design", "Garden development and landscape design" for additional vocational education, advanced training and retraining of qualified personnel of regional companies, authorities and local communities through some advanced technological educational platforms (MOOC, LMS, EdTech, GreenField).

Module 2. Landscape development optimization: creating environmental tourist attractions

2.1. Natural and historical landscapes as objects of local lore tourism (setting up the tourist trust "Balga", "Vishtynets");

2.2. Analysis of old dendrology parks and roadside avenues condition with a view of using them in tourism;

2.3. Mounted bogs are the cores for the natural framework, the habitats of rare and endangered species of animals, plants and fungi of Central Europe;

2.4. Assessment of the key ornithological territories condition with a view to initiating an educational tourism;
2.5. Assessment of the coastal landscapes potential for the development of local tourism.
2.6. Wetlands and their role in maintaining a stable equilibrium of coastal-water landscapes.

Module 3. History of land use and population settlement

3.1. Analysis of land use and landscape treatment in the exclave in the 19th and first half of the 20th century so that the historical experience can be used in modern conditions;
3.2. Monitoring the use of the territories of municipal entities located on anthropogenic (polder) landscapes (Slavsky, Polessky, Guryevsky districts) in historical context to increase their competitiveness.

Module 4. Landscape culture as a tool for developing civil society, raising their awareness, educating and growing a sense of responsibility among the population at large for preserving the natural and historical landscapes on the territory of the Russian exclave

4.1. Mass media and the Internet as the instruments to make natural scientific knowledge more accessible;
4.2. Organization of public hearings at the stage of preparing and implementing economic projects at the regional and municipal levels;
4.3. Learning public opinion about the state of the environment in the exclave territory through public opinion polls;
4.4. Creating a street youth theater "Green planet" which brings together the regional government youth initiatives centers, the Kaliningrad City Hall, the student scientific society of the federal university, the center for environmental education and tourism of the Ministry of Education of the Kaliningrad region.

Module 5. Legislative activity as the background for taking landscape development planning steps

5.1. Initiating the work on by-laws to the Kaliningrad Regional Law "On the Red Book of Endangered Species of the Kaliningrad Region" (Adopted by the Kaliningrad Regional Duma of the fourth convocation on April 22, 2010, brought into force on May 4, 2010, No. 442, Kaliningrad) for the conservation of biological diversity, protection and reintroduction of rare and endangered species (subspecies, populations) of animals, plants and fungi, preservation and restoration of their habitats and growth in the territories of specific municipal entities of the Kaliningrad region, as well as for environmental education, training and promotion of environmental knowledge.

Module 6. Expert assessment and consultancy

6.1. Promotion of the provisions of the Kaliningrad regional landscape program as the background for strategic planning of the territory development and the environmentally oriented use of the region's natural resources through a Resolution of the Governor of the Kaliningrad Region;

6.2. Improvement of the legislation on setting and managing specially protected nature conservation areas at the regional and municipal levels (including the preparation work on the Kaliningrad Regional Law "On assigning lands to specially protected natural areas of regional and municipal significance");

6.3. Further works on setting a regional network of specially protected natural areas (core natural areas) and their efficient exploitation;

6.4. Development and implementation of the regional program of environmentally-focused forest management (including restoration of forest communities on watersheds and river valleys);

6.5. Development of proposals for the rehabilitation and reclamation for the territories which sustained damage due to economic activities: amber and sand-gravel pits; cattle cemeteries, authorized and unauthorized landfills of solid domestic waste (SDW), beds and shores of the most polluted waterways;

6.6. Widespread propaganda of the advantages of cleaning domestic sewage at the communal level in treatment plants of the vegetative-marsh type;

6.7. Working out and implementing a soil protection program in the areas of surface and underground runoff formation;

6.8. Participation in the development and implementation of the regional program of environmental and local.

Module 7. International cooperation with the EU countries (Poland, Lithuania, Germany, Denmark, Sweden) for the conservation of the biotopic diversity of natural and natural—historical transboundary landscapes (the Vishtynets Upland, the Neman River Delta, the Curonian and Kaliningrad Gulfs, the Curonian and Baltic Spits). The ILC's legislative initiative will promote the adaptation and acceptance of a number of basic directives (the "Landscape Directive", the "Water Directive", the "Biodiversity Directive" etc.) adopted in the Council of Europe and the EU countries to the conditions of the Kaliningrad region and in the future to Russian conditions.

Conclusion

1. ILC's partners are federal, regional and municipal authorities responsible for resolving issues in ecology and environmental protection.

2. ILC will be in demand with ministries and departments of the Government of the Kaliningrad region; Kaliningrad Regional Duma; Public Chamber of the Kaliningrad Region; municipalities of the Kaliningrad region.

References

Antipov AN, Drozdov AV (eds) (2002) Landshaftnoe planirovanie: printsipy, metody, evropeiskii i rossiiskii opyt (Landscape planning: principles, methods, European and Russian experience). Institut geografii SO RAN Publ, Irkutsk (in Russ.)

Bundesamt für Naturschutz mit schwerpunkt landschaftsplaning (2006) Deutsch-Russisch-Englisches sachworterbuch. Bonn-Berlin-Hannover-Moskau-Irkutsk

Dedkov VP et al (1999) Rastitel'nost' (Vegetation). In: Berenbein DY (ed) Kaliningradskaya oblast'. Prirodnye resursy. Yantarnyi skaz, pp 139–148 (in Russ.)

Dedkov VP (ed) (2004) Biologicheskie resursy Kaliningradskoi oblasti i puti ikh ratsional'nogo ispol'zovaniya (Biological resources of the Kaliningrad region and ways of their rational use). Kaliningrad State University Publication, Kaliningrad (in Russ.)

Dedkov VP, Fedorov GM (2006) Prostranstvennoe, territorial'noe i landshaftnoe planirovanie v Kaliningradskoi oblasti (Spatial, territorial and landscape planning in the Kaliningrad region). Immanuel Kant Russian State University Publ, Kaliningrad (in Russ.)

Dedkov VP, Grishanov GV (2010) Krasnaya kniga Kaliningradskoi oblasti. Zhivotnye Rasteniya Griby Ekosistemy (The red book of the kaliningrad region. Animals plants mushrooms ecosystems). Kaliningrad: Immanuel Kant Russian State University Publ. (in Russ.)

Litvin VM, El'tsina GN, Dedkov VP (1999) Kaliningradskaya oblast'. Prirodnye resursy (Kaliningrad region. Natural resources). Yantarnyi skaz (in Russ.)

Ministry of Natural Resources and Ecology of the Kaliningrad Region (2017) Gosudarstvenyi doklad «Ob ekologicheskoi obstanovke v Kaliningradskoi oblasti v 2016 godu» (State report on the environmental situation in the Kaliningrad region in 2016). Kaliningrad (in Russ.)

Mistletoe Infestation as a Transboundary Problem

Natalia Y. Chupakhina⬤, Pavel V. Maslennikov⬤,
Liubov N. Skrypnik⬤, Pavel V. Feduraev⬤
and Galina N. Chupakhina⬤

Abstract

The rapid spread of the hemiparasitic shrub of the European white-berry mistletoe, *Viscum album* (L.), in the Kaliningrad region is alarming. One of the probable reasons for this phenomenon is global warming. Another factor favourably affecting the development and reproduction of mistletoe is the increase in the number of bird species feeding on it and dispersing its seeds. The lack of a scientifically based and properly organized control against this phenomenon could be regarded as yet another factor contributing to the spread of mistletoe. Moreover, there is no single opinion on the mistletoe effect on woody plants. Therefore, the question arises of the scrupulous study of biology, physiology, aetiology and biochemistry of mistletoe in the Kaliningrad region, along with the issue of mechanical mistletoe removal. In this connection, the experience of neighbouring countries inhabited by the plant and using it for various purposes is of particular interest.

Keywords

The European white-berry mistletoe (*Viscum album* (L.)) · Spread ·
The Kaliningrad region · Removal

N. Y. Chupakhina
Faculty of Bioresources and Nature Management, Kaliningrad State Technical University, Kaliningrad, Russia

P. V. Maslennikov (✉) · L. N. Skrypnik · P. V. Feduraev · G. N. Chupakhina
Institute of Living Systems, Immanuel Kant Baltic Federal University, Kaliningrad, Russia
e-mail: PMaslennikov@kantiana.ru

© Springer Nature Switzerland AG 2020
G. Fedorov et al. (eds.), *Baltic Region—The Region of Cooperation*,
Springer Proceedings in Earth and Environmental Sciences,
https://doi.org/10.1007/978-3-030-14519-4_16

Introduction

Due to the influence of the climatic and anthropogenic factors, species not characteristic of certain ecotopes are forced to shift their ranges to find a new ecological niche. During the last 10 years, the number of trees infected by the semiparasitic shrub of the European white-berry mistletoe, *Viscum album* (L.), has been increasing in the Kaliningrad region. The European mistletoe is a recognized stressor for host trees and shrubs (Sangüesa-Barreda et al. 2018). Due to the fact that mistletoe is spread by birds, and birds do not recognize administrative borders, the problem of the Kaliningrad region becomes a problem for its neighbouring countries. Thus this issue is of a transboundary nature.

Material

The genera Viscum belonging to the mistletoe family Loranthaceae includes about 100 species of evergreen semiparasitic plants growing in subtropical, tropical and temperate regions of Eurasia. According to host specificity, the mistletoe is classified into three subspecies: *Viscum album* L. Scp. Platyspermum growing on deciduous plants; *Viscum album* L. Scp. Abietis growing only on Abies alba, European silver fir; *Viscum album* L. Scp. Laxum—a pine mistletoe growing on pine trees and rarely on spruce (Vardanyan et al. 2011).

The most common species in western and southern Europe is the European white-berry mistletoe (*V. album* (L.)). In the north and east of Europe, there is *V. coloratum* (Kom.). Its appearance is quite similar to that of the European mistletoe; however, its berries are yellow and orange. Latvia represents the northern range limit of the European mistletoe in the Baltic States. Both species occur in Russia with *Viscum coloratum* (Kom.) found in the Far East. In the countries bordering Russia, coloured mistletoe is a strictly protected species; therefore, its study in recent years has focused primarily on its genetic and ecological conservation (Kim et al. 2017).

Mistletoe is an evergreen semiparasitic shrub epiphyte of spherical shape with a diameter of 20–150 cm growing on the branches of deciduous trees (oak, poplar, willow, maple, birch, linden, pear, elm, apple, hawthorn, walnut, hornbeam, false acacia, plum, pine, spruce, etc.). The most susceptible to mistletoe infestation are trees with increased water balance, hygrophytes and softwood trees. Nevertheless, mistletoe can occur on hardwoods and weakened trees.

Results

Mistletoe is dioecious having separate male and female plants. Through root-like structures called haustoria that penetrate into the wood, mistletoe attaches to a host tree and absorbs its nutrients. Mistletoe is dichotomously branched. It is possible to estimate the age of a mistletoe bush simply by counting the number of forks, as its green and brown branches fork once per year.

The branches are brittle, breaking apart easily at the nodes. The yellowish green leaves are opposite, sessile, coriaceous with 3–4 clearly marked longitudinal veins. The leaves contain chlorophyll that makes mistletoe capable of photosynthesis, thus the shrub is only semiparasitic. Moreover, when a host tree sheds its leaves mistletoe might provide it with carbohydrates. Mistletoe itself sheds in autumn with only second-year leaves falling. It shows a high rate of transpiration ensured by a supplementary mechanism of stomatal opening. Transpiration rate in some mistletoe species is ten times higher than in their host plants. The small, yellowish-green unisexual flowers of mistletoe grow in groups in the top forks of the branches. Staminate flowers are larger than pistillate which do not exceed 2 mm in diameter. Mistletoe fruit is a round pseudo-berry, approximately 10 mm in diameter. Unripe fruit is green, ripe one is translucent white. The seed is large, approximately 8 mm in diameter, embedded in a viscid pulp in a pod coat. Seeds contain up to three embryos.

At the moment, there is no data on the ecological relationships between mistletoe and its host trees in the Kaliningrad region, and the information on the anatomical and morphological characteristics associated with the specific nature of this species is not sufficient.

In our zone, the flowering season is in late winter. White berries remain on the plant for over a year. With its seeds being dispersed by berry-eating birds, mistletoe is an ornithochorous plant. Birds pecking its fruit in autumn and winter spread seeds long distances when flying to other woodlands and migrating to nesting areas. The seed either passes through the bird's digestive system and germinates on tree barks where it falls or adheres to the bird's beak, and is removed while the bird tries to wipe it off by scraping it on to another tree's branch. The seed glues to a new host forming haustoria that penetrate the tree's bark. Viscin contained in mistletoe berries and covering its seeds is a crucial factor in the survival of the species.

The list of birds cited with the European mistletoe fruit consumption includes 28 species (Nechaev 2008). According to the effect their beaks and digestive tracts have on fruit and seeds, mistletoe-eating birds form the following groups (Nechaev 2008):

1. Birds breaking fruit and seeds into smaller pieces with their beaks; mostly consuming "kernel" seeds only (finches: Pine Grosbeak, Hawfinch, Redpoll and others; tits; Tree Sparrow).
2. Birds swallowing whole fruits and grinding them completely or partially with gastroliths in gizzard (Galliformes: Hazel Grouse, Pheasant and others).

3. Birds swallowing whole fruits not damaging seeds with their beaks. They digest
   only the fruit pulp while the seeds remain undamaged and are later defecated
   (Grey-headed and Black Woodpeckers, Grey Starling, Corvidae: Jay,
   Azure-winged Magpie, Nutcracker, Large-billed Crow, Carrion Crow; Dusky
   Thrush and other thrushes; Bohemian Waxwing, Japanese Waxwing).

The enormous increase in the number of representatives of the family Corvidae
in cities, including the urban area of the Kaliningrad region, resulted in these birds'
contribution to the spread of mistletoe. This growth is caused by unorganised
landfill sites around towns and villages providing a plentiful food supply.

Discussion

Mistletoe is a wild-growing medicinal plant. Today, the content of antioxidants in
medicinal plants, particularly in mistletoe, is drawing a lot of attention (Onay-Ucar
et al. 2012). The intensification of free radical oxidation damages the structure and
properties of lipid membranes establishing the direct correlation between the excess
of free radicals and the risk of dangerous diseases (Chupakhina et al. 2014). Plant
antioxidants are biologically active substances that bind excessive free radicals and
inhibit increased lipid oxidation and the formation of unwanted oxidation products
(Chupakhina et al. 2014). In this regard, the evaluation of the antioxidant properties
of plants is of some practical interest, with mistletoe as a medicinal plant being of
particular interest.

The European mistletoe has been used since ancient times (V century BC) and is
still used in traditional medicine for treating a wide variety of diseases:
atherosclerosis, inflammatory kidney disease, chronic metritis, gastric and pancre-
atic diseases, pulmonary tuberculosis, bronchial asthma, hemorrhoids, hemor-
rhages, diabetes mellitus, etc. (Vardanyan et al. 2011). A large number of medicinal
substances in mistletoe makes it possible to use the plant to treat hypertension,
neuralgia, asthma, degenerative disk disease, rheumatism and gout, swollen lymph
nodes and abscesses. Some researchers discuss the issue of using mistletoe for
treating human and animal cancer (Christen-Clott et al. 2010; Zhao et al. 2011). The
reason for such a "universal" therapeutic effect of the European mistletoe is its
complex and rich chemical composition. The leaves and young shoots are used for
medicine production as they contain oleanolic and ursolic acids. They are also rich
in resinous substances and fatty oils (Zarkovic et al. 1998). Other agents present in
mistletoe include alkaloids, saponins, lupeol, viscerin, viscotoxin, choline and
acetylcholine, propionylcholine and tyramine (Zarkovic et al. 1998; Vardanyan
et al. 2011).

Mistletoe leaves also contain carbohydrates, polyhydric alcohols, organic acids,
triterpenoids, gum, sterols, nitrogen compounds, polypeptides, lectins, A, C, E
vitamins, phenols and their derivatives, tannins, phenolcarboxylic acids and
their derivatives, flavonoids, chalcones, higher fatty acids, wax, carotenoids

(Vardanyan et al. 2011). Mistletoe berries contain fatty oil, gum, resinous substances, carotene, and ascorbic acid. The above chemical constituents of the European mistletoe, including flavonoids, carotenoids, vitamins A, C, E, nitrogen-containing compounds, and phenols, indicate its antioxidant properties (Zhao et al. 2011; Nazaruk and Orlikowski 2015). However, the chemical composition of the species Viscum, in particular, *V. album* L., is argued to be not thoroughly studied, and the antioxidant properties of the extract of white-berried mistletoe leaves are of certain interest. The extract of another mistletoe type, *V. coloratum*, contains 41 identified compounds, including 11 flavonoids, 2 hormones, 14 benzenoids, inositol, 2 pyrimidines, 4 triterpenoids, viscolin, 5 steroids, and a new flavanone-(2S)-7,4'-dihydroxy-5,3'-dimethoxyflavanone (Yao et al. 2006, 2007). Recently Chinese researchers have extracted a homoeriodictyol-7-O-Beta-ᴅ-glucopyranoside from *Viscum coloratum*. This flavonoid has an inhibitory effect upon platelet-activating factor causing bronchospasm and eosinophilic infiltration of airway mucosa (Chu et al. 2008; Zhao et al. 2007). This natural inhibitor provides a basis for drug development. Moreover, all flavonoids of *V. album* and *V. coloratum* have antioxidant activity (Yao et al. 2006, 2007) and cardioprotective properties (Chu et al. 2008; Zhao et al. 2007).

Mistletonone, diarylheptanoid from *Viscum coloratum*, and 1,3-diphenylpropane (viscolin) (Su et al. 2006), also have a pronounced antioxidant effect and can be used for treating diseases caused by oxidative stress (Yao et al. 2006, 2007).

Viscotoxins and lectins stand out of known mistletoe components. (Ochocka and Piotrowski 2002). Viscotoxins are the most active protein substances, rich in cysteine, particularly a- and b- thionines having an effect on the plant's membrane. There are four identified mistletoe's viscotoxins: A1, A2, A3 and B (Klein et al. 2002). Their amino acid sequence shows their high homology. Viscotoxins can assemble into complexes with other protein toxins, thereby changing their cytotoxic effect on target cells. Presumably, viscotoxins provide plants with the protection from viruses, bacteria and fungi. There are data (Klein et al. 2002) indicating the stimulating effect of viscotoxins on antibody production. Despite ongoing studies, the function and mechanism of their biological effects remain unclear (Yao et al. 2006).

Being a semiparasite, the mistletoe harms its host plant. Some authors believe that mistletoe infestation results in the death of the tree (Yelpitiforov et al. 2017). Other researchers believe that mistletoe can be used not only as an indicator of the health of woody plants but also as an environment pollution indicator (Yelpitiforov et al. 2017). Heavy-metal sensitive species most correctly reflecting the pollution level in urban areas are of considerable importance as bioindicators of heavy metal contamination (Chupakhina et al. 2017). Mistletoe leaves and haustoria have exceptional heavy metal-accumulating capacity resulting in host plant weakening. Moreover, as mistletoe feeds on its host's mineral elements it absorbs nutrients from the tree as well as stimulates the accumulation of certain elements rather than others, e.g., barium (Yelpitiforov et al. 2017).

Furthermore, mistletoe can be a stress test for wood species' resistance to colonization by this and other semiparasitic plants. This will allow for determining a plant species composition that is resistant to mistletoe infestation in order to prevent it from spreading.

Conclusions

There is no standard technique for *Viscum album* and its subspecies management. However, the only effective control method is the mechanical removal of mistletoe branches from a host tree. Extensive use of mistletoe's active components would promote its removal.

Mistletoe is mechanically harvested to be used as Christmas and wedding decorations. The Kaliningrad region does not have this decorative tradition. However, the plant is widely used to create a festive mood in winter, for example, in Germany. This might be the reason why mistletoe is visually not observed on the trees in Berlin, Dresden, Hanover, Munich or Hesse.

Despite the unusual ornamental appearance of the trees with green mistletoe balls on them, especially when leaves fall, it should be borne in mind that this semiparasite causes tree dehydration and reduces its wind resistance. In the Kaliningrad region mistletoe cut during sanitation pruning is transported to a solid waste site and disposed together with other pruning wastes, although it appears on some regional Red lists as a rare and protected species (Velichkin 2012). Not exploiting the region's potential as the source of this medicinal raw material seems unpractical.

The level of mistletoe infestation in the Kaliningrad region is so high that it demands action. Planting trees that are not susceptible to mistletoe infestation will prevent it from spreading. However, this is a long-term solution. The required immediate intervention is mistletoe mechanical harvesting for medicinal and decorative purposes (floral compositions). Sanitation mistletoe pruning will result in improved host vigour and increased wind resistance, as well as it will have a positive effect on the transboundary spread of this semiparasitic shrub.

Acknowledgments This research was supported by the Russian Academic Excellence Project at the Immanuel Kant Baltic Federal University.

References

Christen-Clott O, Klocke P, Burger D (2010) Treatment of clinically diagnosed equine sarcoid with a mistletoe extract (*Viscum album* austriacus). J VetY Intern Med 24:1483–1489. https://doi.org/10.1111/j.1939-1676.2010.0597.x

Chupakhina GN, Maslennikov PV, Skrypnik LN, Chupakhina NYu, Poltavskaya RL, Feduraev PV (2014) The influence of the Baltic regional conditions on the accumulation of water-soluble antioxidants in plants. Russ Chem Bull 63:1946–1953. https://doi.org/10.1007/s11172-014-0684-6

Chupakhina GN, Maslennikov PV, Mosina LV, Skrypnik LN, Dedkov VP, Chupakhina NYu, Feduraev PV (2017) Accumulation of biogenic metals in the plants of urbanised ecosystems in the city of Kaliningrad. Res J Chem Environ 21:9–17

Chu W, Qiao G, Bai Y, Pan Z, Li G, Piao X, Wu L, Lu Y, Yang B (2008) Flavonoids from Chinese *Viscum coloratum* produce cytoprotective effects against ischemic myocardial injuries: inhibitory effect of flavonoids on PAF-induced Ca^{2+} overload. Phytother Res 22:134–137. https://doi.org/10.1002/ptr.2267

Kim BY, Park H, Kim S, Kim YD (2017) Development of microsatellite markers for *Viscum coloratum* (Santalaceae) and their application to wild populations. Appl Plant Sci 5:1600102. https://doi.org/10.3732/apps.1600102

Klein R, Classen K, Fischer S, Errenst M, Scheffler A, Stein GM, Scheer R, von Laue HB (2002) Induction of antibodies to viscotoxins A1, A2, A3, and B in tumour patients during therapy with an aqueous mistletoe extract. Eur J Med Res 7:359–367

Nazaruk J, Orlikowski P (2015) Phytochemical profile and therapeutic potential of *Viscum album* L. Nat Prod Res 30:1–13. https://doi.org/10.1080/14786419.2015.1022776

Nechaev VA (2008) Ob ekologicheskikh svyazyakh mezhdu ptitsami i omeloi okrashennoi *Viscum coloratum* v Primor'e i Priamur'e (On ecological relations between birds and mistletoe stained *Viscum coloratum* in Primorye and Amur Region). Rus Ornitol-Cheskii Zhurnal 408:443–447 (in Russ.)

Ochocka JR, Piotrowski A (2002) Biologically active compounds from European mistletoe (*Viscum album* L.). Can J Plant Pathol 24:21–28. https://doi.org/10.1080/07060660109506966

Onay-Ucar E, Erol O, Kandemir B, Mertoglu E, Karagoz A, Arda N (2012) *Viscum album* L. Extracts protects HeLa cells against nuclear and mitochondrial DNA damage. Evid-Based Complement Altern Med 1:958740. https://doi.org/10.1155/2012/958740

Sangüesa-Barreda G, Camarero J, Pironon S, Pelegrín E, Gazol A, Peguero-Pina J, Gil-Pelegrín E (2018) Delineating limits: confronting predicted climatic suitability to field performance in mistletoe populations. J Ecol. https://doi.org/10.1111/1365-2745.12968

Su CR, Shen YC, Kuo PC, Leu YL, Damu AG, Wang YH, Wu TS (2006) Total synthesis and biological evaluation of viscolin, a 1,3-diphenylpropane as a novel potent anti-inflammatory agent. Bioorganic Med Chem Lett 16:6155–6160. https://doi.org/10.1016/j.bmcl.2006.09.046

Vardanyan RL, Vardanyan LR, Atabekyan (2011) Antioxidant effect of extracts of white mistletoe (*Viscum album* L.) grown in different trees. Chem J Armen 64:335–343

Velichkin EM (2012) O rasprostranenii Omely beloi (*Viscum album* L, Lorenthaceae) v Bryanskoi oblasti (On the distribution of mistletoe whites (*Viscum album* L, Lorenthaceae) in the Bryansk region). Vestn Bryanskogo Gos Univ 4:124–126

Yao H, Liao Z-X, Wu Q, Lei G-Q, Liu Z-J, Chen D-F, Chen J-K, Zhou T-S (2006) Antioxidative flavanone glycosides from the branches and leaves of *Viscum coloratum*. Chem Pharm Bull 54:133–135. https://doi.org/10.1248/cpb.54.133

Yao H, Zhou GX, Wu Q, Lei GQ, Chen DF, Chen JK, Zhou TS (2007) Mistletonone, a novel antioxidative diarylheptanoid from the branches and leaves of *Viscum coloratum*. Molecules 12:312–317. https://doi.org/10.3390/12030312

Yelpitiforov EM, Ivanytska BA, Malashuk OV (2017) Comparative analysis of the content of chemical elements *Viscum Album* L. and *Viscum Album* Subsp. Austriacum (Wiesb.) Vollmann. Scientific Bulletin of UNFU 27:93–97. https://doi.org/10.15421/40270519

Zarkovic N, Kalisnik T, Lonöaric I, Boovic S, Mang S, Kissel D, Konitzer M, Jurin M, Grainza S (1998) Comparison of the effects of *Viscum album* lectin ML1 and fresh plant extract (Isorel) on the cell growth in vitro and tumorigenicity of melanoma B16F10. Cancer Biother Radiopharm 13:121–131. https://doi.org/10.1089/cbr.1998.13.121

Zhao Y, Wang X, Zhao Y, Gao X, Bi K, Yu Z (2007) HPLC determination and pharmacokinetic study of homoeriodictyol-7-O-beta-D-glucopyranoside in rat plasma and tissues. Biol Pharm Bull 30:617–620. https://doi.org/10.1248/bpb.30.617

Zhao Y, Yu Z, Fan R, Gao X, Yu M, Li H, Wei H, Bi K (2011) Simultaneous determination of ten flavonoids from *Viscum coloratum* grown on different host species and different sources by LC-MS. Chem Pharm Bull 59:1322–1328. https://doi.org/10.1248/cpb.59.1322

An Assessment of Landscape Vulnerability in the Kaliningrad Region: Towards Better Municipal Nature Management

Ivan I. Kesoretskikh⑩, Sergeyi I. Zotov⑩ and Natalia N. Lazareva

Abstract

In this article, we present a technique for assessing landscape vulnerability to human impacts at a municipal level. The technique was tested in the Pravdinsk district of the Kaliningrad region. Our aim was to adapt the assessment technique for municipal use. Our landscape map of the study area, as well as relevant topographic maps, comprised the cartographic framework for the vulnerability assessment. An integral matrix of parameters and weighting factors for the study area was instrumental in creating a scheme of landscape areas most vulnerable to human impacts. The scheme obtained was juxtaposed with the Pravdinsk municipality's urban planning documents. A comparative analysis of these documents proved local spatial planning to be sporadic and assessments of the environment and of environmental priorities absent. In view of the observed landscape vulnerability differentiation, we prepared proposals for the location of engineering infrastructure and industrial facilities. Our study shows that our technique for assessing landscape vulnerability to human impacts and its integration into the system of municipal nature management will contribute to a balanced assessment of municipalities' geoecology—a *sine qua non* of spatial planning.

Keywords

Landscape vulnerability assessment technique · Human impact · Municipal nature management · Spatial planning

I. I. Kesoretskikh
Institute for Spatial Planning Development and External Relations, Kaliningrad, Russia

S. I. Zotov (✉) · N. N. Lazareva
Institute of Environmental Management, Urban Development and Spatial Planning, Immanuel Kant Baltic Federal University, Kaliningrad, Russia
e-mail: zotov.prof@gmail.com

© Springer Nature Switzerland AG 2020
G. Fedorov et al. (eds.), *Baltic Region—The Region of Cooperation*,
Springer Proceedings in Earth and Environmental Sciences,
https://doi.org/10.1007/978-3-030-14519-4_17

153

Introduction

The search for tools and techniques for integrating sustainable approaches into the Russian spatial planning system is a much-discussed interdisciplinary problem.

According to the current Urban Planning Code of the Russian Federation, spatial planning stands for 'planning the development of territories, in particular, for the purposes of functional zoning, object location in line with public and municipal needs, and the designation of restricted use areas' (Gumenyuk 2016). It has been noted (Fedorov et al. 2015) that planning spans the federal, regional and municipal governance levels, the latter being responsible for municipal spatial planning schemes and urban and rural master plans. The keystones of spatial planning in Russia are sustainable development (paragraph 1, article 2 of the Urban Planning Code of the Russian Federation) and regard to environmental, economic, social, and other factors (paragraph 2, article 2 and article 9) (Urban Development Code of the Russian Federation 2017).

Thus, municipalities have to create conditions for a balanced spatial development in line the architecture of regional nature management. Finding a solution to this complicated problem requires an ecological status assessment, an analysis of the distribution of human impact sources, etc. To this end, one may employ techniques used in comprehensive geo-ecological analyses, for instance, that for assessing natural system vulnerability to human impacts. In European countries, this approach has been effectively applied for decades. Moreover, it has been included in the concept of municipal landscape planning (Golobič and Breskvar 2010). All this has resulted in a shift to a multifactor assessment of a study area's environmental condition, which made it possible to identify existing and potential conflicts in nature management.

The Object of Research and Methodology

The study area is located in the Pravdinsk district in the north of Russia's Kaliningrad region (Fig. 1). The municipality borders on the Republic of Poland.

The study area was not chosen randomly. In terms of environmental protection, the site in question has not only regional but also international significance. The municipality's territory is drained by the tributaries of the River Zhernovka, which flows into the River Łyna. Lying on both sides of the national border, the river's catchment area offers an attractive recreational setting. However, it is classed as a territory, the natural systems of which are very sensitive to human occupation.

The study area spans 23.4 km². A landscape map of the study area and relevant topographic maps comprised the cartographic framework for the research.

This study is a search for ways to optimise municipal nature management by using comprehensive geo-ecological assessment data and employing a technique for multicriteria assessment of natural system vulnerability to human impacts. Earlier, such a technique was tested and proved effective in regional-level assessments (Zotov et al. 2013). The technique involves the following stages:

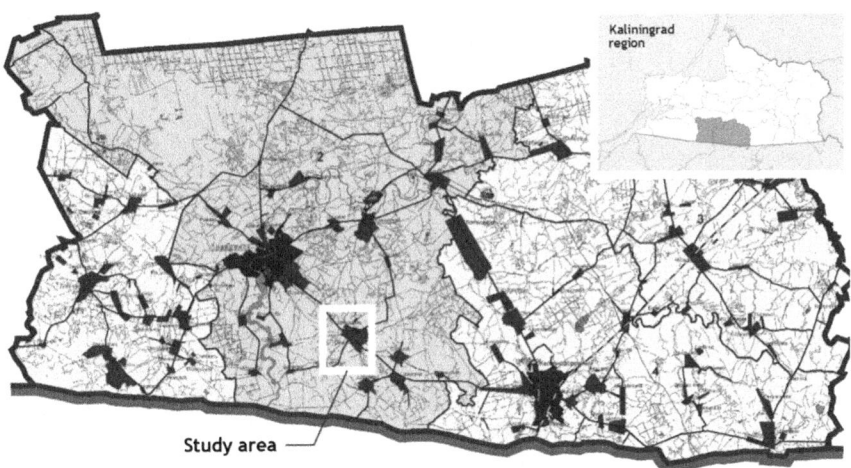

Fig. 1 The study area

1. The selection and justification of vulnerability assessment criteria;
2. The creation of a parametric matrix and the grading of assessment criteria according to vulnerability classes;
3. The calculation of weighting factors for vulnerability assessment parameters;
4. The selection of an optimal territorial unit for landscape vulnerability assessment.

When selecting the parameters, we took into account the following factors:

1. The main functional elements of a landscape are the hydrological cycle, geo-chemical cycles, and energy exchange. One of the leading functions of a landscape is supporting the flow of substances and energy;
2. The main features of a landscape's structure correlate with a range of abiotic indicators;
3. The assessment parameters should take into account the characteristics of the object of research and be in line with the objectives of the study;

To solve the above problems, we created a matrix (Table 1) that takes into account the environmental conditions of the Kaliningrad region.

The distance to a watercourse (m) is the shortest distance from the source of impact or from the study area to the nearest surface watercourse. In an integral formulation, this parameter describes the subordination of landscapes that determine pollutant migration and accumulation.

The spawning status of a watercourse identifies whether a watercourse—or any of its parts—is classed as a restricted use area.

The conservation status determines whether the territory under consideration—or any of its parts—is classed as a restricted use area.

Table 1 A matrix of landscape vulnerability parameters

Parameter	Vulnerability class									
	High		Increased		Moderate		Reduced		Low	
	From	To	From	To	From	To	From	To	From	To
Distance to a watercourse (m)	0	200	201	400	401	600	601	800	801	1000
Slope (°)	20	17	16	13	12	9	8	5	4	0
River network density (km/km^2)	1.4	1.25	1.24	1.11	1.10	0.96	0.95	0.80	0.79	0.60
Spawning status	Yes				No					
Conservation status	Yes				No					
Water table (m)	0.5	2.0	2.1	4.0	4.1	6.0	6.1	8.0	8.1	10.0
Soil texture	Sand		Loamy sand		Sandy clay loam		Sandy clay		Clay	
Land type	Swamp		Forest				Grassland (agricultural lands)			

Source Kesoretskikh et al. (2015)

The water table (m) stands for the depth of the aquifer that is closest to the surface. Groundwater is considered a channel for substance and energy transport within landscapes.

The soil texture, in an integral formulation, stands for the level of chemical pollutant infiltration and describes the probability of erosion and slope degradation processes.

Land type stands for the land use category—swamp, forest, or agricultural grassland.

River network density (km/km^2) is the ratio of the length of all the local surface watercourses to the area of the territory in question. The higher this value is the more active is the migration of pollutants through the surface watercourses.

The slope (°) is the incline of the study area. This parameter is important for runoff generation (migration of chemical pollutants) and erosion processes.

We calculated weighting factors based on 20 parameter significance distribution scenarios. Based on the data obtained, two groups of parameters were identified—primary (distance to the watercourse, water table, soil texture) and secondary (spawning and conservation status, slope, river network density, land type). The primary factor was set at 0.25 and the secondary at 0.05.

To obtain an optimal territorial unit, we used a grid with 1 km grid spacing.

In our calculations, we employed the 'Assessment of landscape vulnerability to human impacts in the Kaliningrad region' GIS, which we had developed using the ESRI ArcGIS software.

The technique made it possible to produce a regional scheme of landscape areas most vulnerabile to human impacts.

To adapt the proposed technique for a municipal level, we used a grid with 0.25 km grid spacing to obtain an optimal territorial unit. The use of such a unit is justified by the size of the study area and the pattern of assessment criteria distribution. The study employed the above integral matrix of parameters and weighting factors calculated using the aggregated indices randomization method (Kesoretskikh et al. 2015). As a result, we obtained a scheme of landscape areas most vulnerable to human impacts in the studied area (Fig. 2).

This scheme was juxtaposed with the current spatial planning documents for the Pravdinsk municipality (Trukhachev et al. 2011)—the municipal draft master plan and the spatial planning scheme. The latter included spatial planning proposals, a restricted use area map, and a scheme of a comprehensive analysis of territorial development.

The priority of the Pravdinsk municipal regulation on spatial planning (Administration of the Pravdinsk Municipal District of the Kaliningrad Region 2011) is to ensure sustainable development until 2030 by planning that will take into account social, economic, environmental and other factors. Official documents emphasise the need for sustainable nature management. However, a comparative analysis of spatial planning schemes and most vulnerable plot maps revealed the weaknesses of the current approach, which hampers the creation of a mechanism for a comprehensive environmental assessment and the identification of environmental priorities (NIIPG Radostroitelstva 2014). This circumstance ultimately complicates balanced and effective decision-making as regards municipal conservation measures (Karnatak et al. 2007; Meng et al. 2011) and can potentially lead to conflicts in nature management.

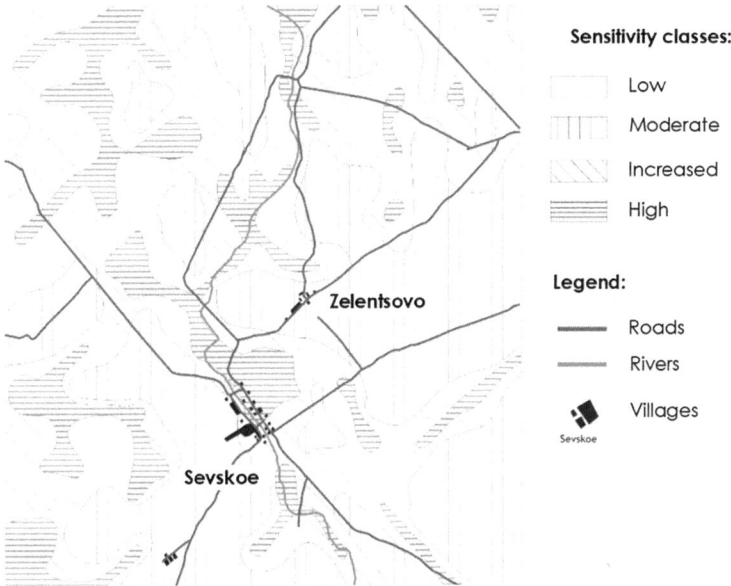

Fig. 2 A scheme of landscape vulnerability to human impacts

By juxtaposing our scheme with the 'Spatial Planning Proposal for the Pravdinsk municipality' (Fig. 3), we identified residential development and engineering infrastructure objects planned for construction in the areas of high landscape vulnerability. One of such areas lies south of the village of Sevskoe—an emerging residential area that will accommodate an inter-settlement gas pipeline and a gas control point.

If it proves impossible to move the objects to lower vulnerabile areas, their construction will require careful attention to the environmental characteristics of the area and a package of conservation measures with a possibility of monitoring the geo-ecological condition of landscapes suffering human impacts.

By juxtaposing a map of restricted use areas with a model of vulnerable landscape areas (Fig. 4), we identified yet another weakness of current approaches to spatial planning, namely, their sporadic nature. Conservation is reduced to the designation of restricted use areas—water protection areas, buffer zones, etc. Moreover, the environmental value of the rest of the territory—outside these zones —is virtually ignored. Little attention is paid to its contribution to the environmental structure of the surrounding natural systems.

The juxtaposition made it possible to identify two most indicative plots. Although not marked on a map of restricted use areas, they belong to high vulnerability areas according to our estimates. Neither transport nor engineering infrastructure nor production facilities should be constructed in these areas, since potential human impacts can damage the elements of local natural systems. Moreover, in some cases (for instance, in those of chemical pollution), negative human impact can spread over vast territories.

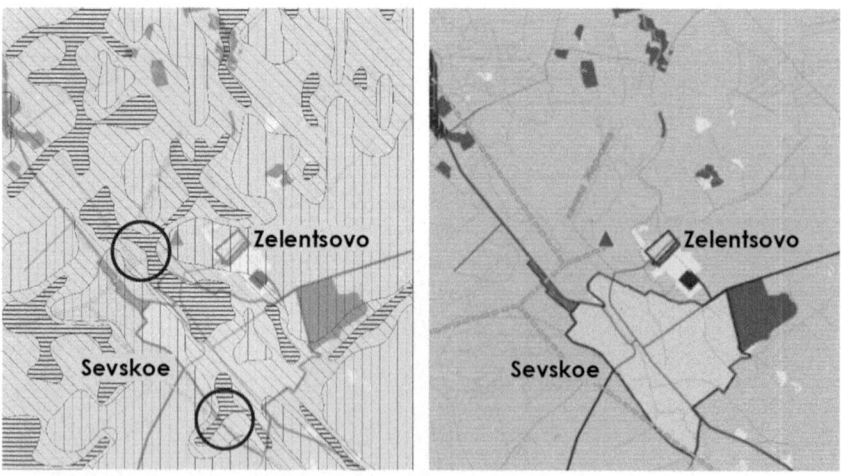

Fig. 3 Our scheme juxtaposed with the 'Spatial planning proposal for the Pravdinsk municipality'

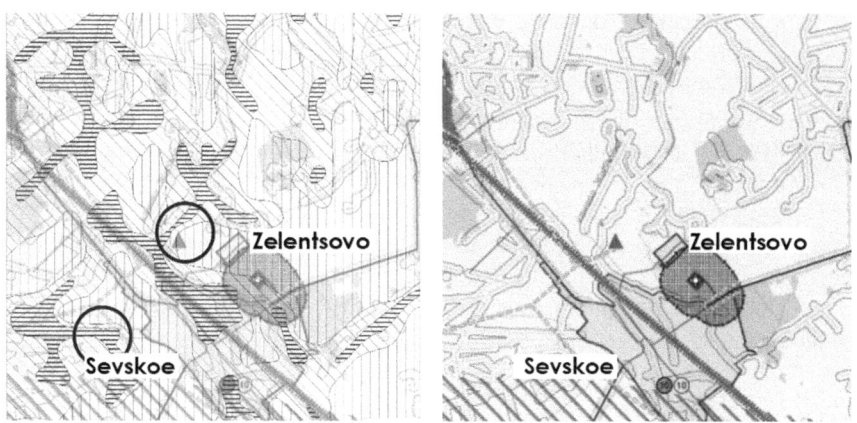

Fig. 4 Our scheme juxtaposed with a map of restricted use areas juxtaposed with

The above examples prove the feasibility of using a comprehensive geo-ecological assessment technique at the local level in optimising municipal nature management. The resulting materials contribute to better conditions for decision-making at the local government level and encourage the use of a multi-criteria approach (Chakhar and Mousseau 2007). In practical terms, the assessment technique is simply a tool for obtaining data necessary for subsequent analysis and decision-making. As a rule, a comprehensive assessment of a territory is preceded by a more ambitious socioeconomic action—an assessment of the conditions for a settlement's development, the selection of infrastructure construction sites, etc. International authors refer to such challenges as 'spatial' problems (Greene et al. 2011). These issues have a pronounced spatial dimension, involve natural objects, and have clear geographic boundaries. Today, such problems affect the interests of different groups. At a municipal level, this process involves local residents, business communities, and authorities. As a rule, conflicts of interests arise immediately when future changes are mentioned. The proposed multicriteria approach is tailored to solve this problem. In particular, it helps to select the optimal development option in view of not only the officials' opinions but also a territory' environmental potential. Note that the approach itself is based not on a single calculation method but rather on an analysis of alternatives (scenarios). A good example thereof is juxtaposing the data of a comprehensive geo-ecological assessment with maps of restricted use areas in the Pravdinsk municipality.

Standards for designating buffer zones of industrial and transport facilities have been already changed once in the Kaliningrad region. This raises the logical question of whether one can call any plans 'long-term' when they rely on regulations that can be revised or changed over the next few years. How can the continuity of decisions be ensured? Our technique for assessing natural landscape vulnerability makes it possible to include natural (landscape) components into decision-making, which gives the process a long-term perspective (Estoque 2011).

In decision-making, it is vital to consider the availability of information to target groups. In the case of municipalities, such a group consists of local administration employees. Thus, there is no need for a component analysis of local natural processes. An optimal tool is an integrated map that encompasses a wide range of parameters and demonstrates the assessment results. This is exactly the problem that we set out to solve by introducing the technique for natural system vulnerability assessment. Overall, the recipient has to understand where the construction of new infrastructure is possible, and where negative environmental impacts will be negligible or even dramatic. The data obtained encourage further dialogue with the other stakeholders—for instance, local residents—during public consultations seeking alternative solutions.

Conclusion

The technique for assessing landscape vulnerability to human impacts and its integration into the system of municipal nature management can be transformed into a balanced assessment of a territory's geo-ecological characteristics. This will contribute to better urban development schemes and residential and industrial development plans, as well as thought-through transport and engineering infrastructure.

Current approaches to zoning and designating restricted use areas are divorced from an understanding of the local environment. As a rule, they focus on individual components and do not consider the entirety of intra-landscape connections.

A juxtaposition of the results of landscape vulnerability assessment and maps of future residential development in the Pravdinsk municipality demonstrated that the latter were fragmentary and did not meet sustainable development requirements. This systemic problem has to be solved at the regional and federal levels by integrating the landscape approach into the spatial planning system.

The adoption of the landscape approach will make it possible to demonstrate and assess development alternatives at different planning levels, before a decision on a certain construction project is made. Another advantage of the approach is that alternative construction locations can be selected based on precise calculations. Overall, when conservation aspects are taken into account and advanced GIS tools are used, the decisions ensuring the sustainable development of municipalities can be made.

References

Administration of Pravdinsky Municipal District of Kaliningrad Region (2011) General'nyi plan munitsipal'nogo obrazovaniya Pravdinskoe gorodskoe poselenie Kaliningradskoi oblasti (General plan of municipal unit Pravdinskoe urban settlement of the Kaliningrad region. Materials on the substantiation of the draft master plan. Rostov-on-Don). http://pravdinsk39.ru/pravdinskoe-gorodskoe-poselenie.html. Accessed 29 Aug 2017 (in Russ.)

Chakhar S, Mousseau V (2007) Spatial multicriteria decision making. LAMSADE University of Paris Dauphine, France. http://www.lamsade.dauphine.fr/mcda/biblio/PDF/ChakharMousseau Inbook2007b.pdf. Accessed 13 June 2014

Estoque RC (2011) GIS-based multi-criteria decision analysis (in Natural resource management). University of Tsukuba. http://giswin.geo.tsukuba.ac.jp/sis/tutorial/GIS-based%20MCDA%20_ RCEstoque.pdf. Accessed 12 June 2014

Fedorov GM et al (2015) Prostranstvennoe planirovanie kak instrument koordinatsii razvitiya portov i gavanei Kaliningradskogo/ Vislinskogo zaliva (Spatial planning as an instrument for coordinating the development of ports and harbors of the Kaliningrad/Vistula Gulf). Kaliningrad (in Russ.)

Golobic M, Breskvar LZ (2010) Landscape planning and vulnerability assessment in the Mediterranean. Regional Activity Centre for the Priority Actions Programme, Ljubljana

Greene R, Devillers R, Luther JE, Brian G (2011) GIS-based multiple-criteria decision analysis. Geography compass 5/6, Blackwell Publishing Ltd., pp 412–432

Gumenyuk IS (2016) O sootnoshenii ponyatii prostranstvennoe, strategicheskoe i territorial"noe planirovanie v rossiiskoi federatsii v konteksterazvitiya regiona kaliningradskogo/vislinskogo zaliva (On the relationship of the concepts of spatial, strategic and territorial planning in the Russian Federation in the context of the development of the Kaliningrad/Vistula Gulf region). IKBFU's Vestnik. Ser Nat Med Sci 1:37–44 (in Russ.)

Karnatak HC, Saran S, Bhatia K, Roy PS (2007) Multicriteria spatial decision analysis in web GIS environment. Geoinformatica. Springer, pp 407–429

Kesoretskikh II, Zotov SI, Drobiz MV (2015) Otsenka prostranstvennoi i vremennoi izmenchivosti pokazatelya uyazvimosti landshaftov Kaliningradskoi oblasti kak komponent ekologicheski orientirovannogo territorial'nogo planirovaniya (Estimation of the spatial and temporal variability of the vulnerability index of landscapes of the Kaliningrad Region as a component of ecologically oriented spatial planning). Baltic Reg 4(26):162–180 (in Russ.)

Meng Y, Malczewski J, Boroushaki S (2011) GIS-based multicriteria decision analysis approach for mapping accessibility patterns of housing development sites: a case study in canmore, Alberta. J Geogr Inf Syst 3:50–61

NIIPG Radostroitelstva (2014) Implementation of environmental principles in the territorial planning of Russia. Report on Phase 3 of the EcoRus Project. http://www.ioer.de/fileadmin/ internet/IOER_Projekte/EkoRuss. Accessed 1 Nov 2014

Trukhachev YN et al (2011) Skhema territorial'nogo planirovaniya Kaliningradskoi oblasti ... munitsipal'nogo obrazovaniya «Pravdinskii raion» Kaliningradskoi oblasti» (Scheme of territorial planning of the municipal unit «Pravdinsky district» of the Kaliningrad region). LLC «Scientific and Project Organization» Southern Town-Planning Center», Rostov-on-Don. https://gov39.ru/vlast/agency/aggradostroenie/zip/territory/stp_pravdinskiy_mr/001_ pravdinskiy_mr_potp.pdf. Accessed 29 Aug 2017 (in Russ.)

Urban Development Code of the Russian Federation of 29.12.2004 N 190-FZ. http://www. consultant.ru/document/cons_doc_LAW_51040/. Accessed 30 Aug 2017 (in Russ.)

Zotov SI, Pokrovsky AV, Kesoretskikh II, Zotov IS (2013) Znachimost" rel'efa dlya otsenki uyazvimosti prirodnykh kompleksov k antropogennym vozdeistviyam (The importance of the relief for assessing the vulnerability of natural complexes to anthropogenic impacts). IKBFU's Vestnik. Ser Nat Med Sci 1:318–322 (in Russ.)

Current Problems in the Analysis of Full-Scale Geoecological Data for the South-Eastern Baltic

Pavel P. Chernyshkov

Abstract

A dramatic increase in full-scale data requires new methods of data analysis and modelling. The example of such data is the sea and land surface satellite remote sensing data obtained during the environmental monitoring in the South-Eastern Baltic region carried out by researchers from Poland, Russia and Lithuania. The paper suggests a new approach to data analysis, which in contrast to the regression statistical analysis takes into account the interaction between a multitude of factors determining the most important parameters. The result is a simple mathematical model that enables to adapt established links between factors of different nature to larger space-time scales taking the changes into account. At the same time, it allows for using qualitative assessments (based on the experts' knowledge and experience in this field of research) as the controlling parameters of the model. At the qualitative level, the study demonstrates the effectiveness of this approach in diagnosing and forecasting eutrophication of the Baltic Sea, as well as its dependence on environmental conditions. The research outlines the applicability of this approach for joint analysis of changes in ecological conditions of offshore and coastal areas.

Keywords

Mathematical modelling · Simulation models · Eutrophication · The Baltic Sea

P. P. Chernyshkov (✉)
Institute of Environmental Management, Urban Development and Spatial Planning,
Immanuel Kant Baltic Federal University, Kaliningrad, Russia
e-mail: PCHernyshkov@kantiana.ru

© Springer Nature Switzerland AG 2020   163
G. Fedorov et al. (eds.), *Baltic Region—The Region of Cooperation*,
Springer Proceedings in Earth and Environmental Sciences,
https://doi.org/10.1007/978-3-030-14519-4_18

Introduction

In recent decades, increased marine activities have exacerbated the problems of scientific support for marine resources management and sustainable development of coastal areas. At the same time, the quality requirements for environmental changes forecasts have increased. This resulted in the implementation of national and international geoecological monitoring programs.

Currently the main purpose of geo-ecological monitoring is to improve the quality of the applied models and management decisions. This is achieved by creating mathematical models of natural processes and phenomena using full-scale data.

The amount of full-scale data on environmental conditions is constantly increasing due to new sources of their acquisition. These are the Earth's surface remote sensing data. Therefore, quality improvement in scientific support requires an effective technology for assimilating the results of full-scale observations aimed at parametrization of mathematical models used to assess and forecast ecological processes. This process is especially important for the territories that are of interest for neighbouring countries. One of these territories is the Baltic Sea being of direct interest to nine countries located on its shores. This approach can be successfully applied to the study of the South-Eastern Baltic with its complexity arising from its transboundary nature and intensity of marine and coastal resource management.

Main Body

The most generally formulated principle of mathematical modelling of environmental processes sounds as follows: the model must be sufficiently complex to describe the process under examination adequately and simple enough for understanding the model's performance.

Of all the classes of models used in scientific research, simulations are most closely corresponding to this principle (Novikov and Novikov 2011).

The work "Adaptive environmental assessment and management" under the editorship of C. S. Holing formulated the principle in reliance to ecological system modelling (Kholling 1981).

Simulations are models that include the existing knowledge of the studied natural system as equations, logical relationships, and expert evaluations. They reproduce the behaviour of an object by means of depicting elementary phenomena and processes retaining their logic and sequence. This allows for using the results of full-scale observations of individual processes to obtain integrated data on the object condition at a specific moment in time and to assess its future changes.

In terms of interrelated processes study, simulation modelling is the best method for understanding complex interactions between processes of different nature and origin. In addition, the method can be employed to analyse the relative sensitivity of the system to changes in its components.

Simulation results can be easily compared with the actual values of variables reflecting natural system processes. When the interference patterns of processes in the modelled system are well known, even minor changes in the input control parameters produce tangible output results.

Historical experimental data and existing notions can be used for hindcasting, i.e. reproducing the past behaviour of the system based on a certain set of model parameters. However, a model adequately representing natural processes within a certain period of time does not always correctly reflect changes in the processes outside this period. Nevertheless, the results of multiple calculations and their comparison with full-scale data enable the adjustment of the control parameters of both the whole model and its individual blocks (for block models) leading to forecast quality improvement.

The most vulnerable area of ecosystem simulation is the one where these models are needed most, that is the management forecasting (Platt et al.). Model construction brings a problem of system flow identification. Researches have paid little attention to system flows; therefore, specialists struggle to find analogies in incomplete and poorly systematized literature sources on the subject or are forced to make speculative guesses.

There are other problems associated with the block-type simulation models. One of them concerns the correct definition of variables and the number of model blocks. Unreasonable aggregation of the model accompanied by a lack of data on its components interaction leads to inadequate results. The second problem is the time scale selection. Relatively simple versions of ecosystem models are homogeneous in time meaning that parameters characterizing the relationship between the elements of the system are defined as constants in a given period of time (month, season, year, years).

To address these problems, it is proposed to use the so-called adaptive approach. It involves the pre-selection (basing on multiple calculations) of such a combination of interference parameters of the studied processes that most accurately recreates the observed changes in each process influencing environmental conditions. The simulation model describes the state of a system on the basis of the model's own properties identified using numerical analysis of existing full-scale data and notions of the interrelations between the processes. Prior to each calculation, the interference parameters can be changed in order to obtain the results closest to the observed values.

Adaptive approach to model investigation allows for using it to determine the system's sensitivity to changes in its elements interactions.

For researches starting with the study of the characteristics of individual processes, and then proceeding to interrelated processes study, simulation modelling is the best method for understanding complex interaction systems. Model dimensionality should be sufficient to reconstruct the key features of the modelled object. However, the model should not contradict the fundamental laws of physics, chemistry, and biology.

Climate changes observed in recent decades create new environmental conditions. To model them the connections between different processes and the patterns of environmental changes can no longer be used (Walther et al. 2002).

In recent years, one of the greatest environmental problems of the South-Eastern Baltic is eutrophication of certain water areas (Carstensen 2013). The processes have especially intensified in recent years, probably due to the ongoing climatic changes (Störmer 2011). As a result, the characteristics of the entire Baltic Sea are changing (Bukanova et al. 2013). To monitor eutrophication, the data of satellite remote sensing of the sea surface are widely used (Bukanova and Stont 2010).

The simplest simulation used as an example is a model for assessing the influence of physical factors (wind speed and direction, water temperature, stability and intensity of vertical circulation), interacting and affecting the development of eutrophication processes. This model allows forecasters to create a computing scheme and to test the validity of the existing notions of the interaction between various parameters affecting the development of eutrophication in the South-Eastern Baltic.

The sequence of actions for creating a model is as follows. The first step is selecting a set of variables x_i, which are relevant to the problem or phenomenon under consideration. Then, they are normalized, meaning the upper and lower bound of each variable x_i is determined. A unit of real time scale is set along with the total number of model time units.

After the selection and normalization of the variables, an interaction matrix (α-matrix) is constructed using the expert evaluations of analysts and scientists having relevant experience and qualifications. Each variable appears in the α-matrix twice, once as a column j and once as a row i. The value of each matrix element is between -1 and $+1$ depending on the degree of its influence. The input value of the matrix α_{ij} (interaction coefficient) corresponding to j-column and i-line represents the first-order effect of the variable x_j on a variable x_i per time unit. This number is either positive or equal to zero, depending on whether the increase in the variable x_j leads to an increase in the variable x_i, or to its reduction, or it remains unchanged. Similarly, it is possible to construct the second matrix (β-matrix), in which the interaction coefficients b_{ij} determine the degree of effect of the variable x_i to a variable x_i (in other words, $d(x_i)d(x_j)$). After this, each of the variables is assigned a certain initial value.

It is not uncommon that the system has variables that have an effect on other variables being independent themselves. Only the column corresponds to such variables in the matrix. An example of a set of variables and α-matrix is shown in Table 1.

Table 1 α-matrix example

Object	Affected subject			
	Wind	сзагрзн.в-в.	BOD	Eutrophication
Wind	1	−0.5	−0.5	−0.5
сзагрзн.в-в.	0	1	1	1
BOD	0	0	1	0
Eutrophication	0	0	1	1

It is followed by the calculation of a parameter that is a quotient of division of the sum of positive effects on the process or system by the sum of the negative impacts (Eq. 1).

$$\varphi_i(T) = \frac{1 + \Delta t |sum\_of\_positive\_effects\_on\_x_i|}{1 + \Delta t |sum\_of\_negative\_effects\_on\_x_i|}$$

$$\varphi_i(T) = \frac{1 + \frac{\Delta t}{2} \sum_{j=1}^{m} \left[|a_{ij} + B_{ij}| - (a_{ij} + B_{ij}) \right] x_j(T)}{1 + \frac{\Delta t}{2} \sum_{j=1}^{m} \left[|a_{ij} + B_{ij}| + (a_{ij} + B_{ij}) \right] x_j(T)} \tag{1}$$

where a_{ij} is an element of the interaction matrix determining the effect of a variable x_j on another variable x_i;

b_{ij} is an element of the derivative matrix determining the effect of a derivative $d(\ln x_j)/dt$ on a variable x_i;

where $T = k\Delta t$ for a positive integer k, and Δt time interval, and

where $B_{ij} = b_{ij} d[\ln x_i(T)]/dt$;
    m is a number of variables in the column;
    Parameter $\varphi_i(T)$ reflects positive, negative or zero changes in the process in the next time step. If negative impacts exceed positive, the index (exponent φ_i) is less than 1, the variable x_i decreases. If there are less negative impacts than positive ones, the φ_i index is more than one, the variable x_i increases.
    The analysis of the results obtained and their comparison with the actually observed parameters allows for changing the coefficients of interaction matrices for recalculations. Once after a number of iterations the satisfactory agreement between the estimated and actual values is achieved, it becomes possible to determine the applicability of the obtained model to diagnose and forecast the studied environmental process.

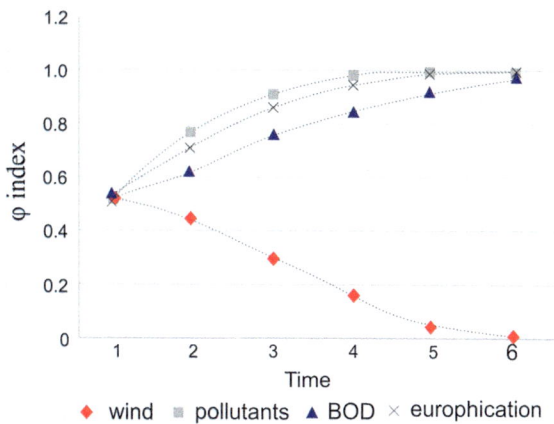

Fig. 1 Interference between various indicators and wind load

The result is a range of simulation indicators, a number of imitative indicators clearly showing the dependency relationships between factors. The graph in Fig. 1 shows that a decrease in wind activity will lead to an increase in BOD, pollution and eutrophication.

Conclusion

The use of the simplest simulation model for analyzing the complex interaction of factors contributing to the emergence and development of the eutrophication process appears to be a promising approach to the analysis of geoecological conditions in coastal areas. Full-scale data allow for the model to "filter" factors that do not have any impact on the environment, contributing to the understanding of the scale and mechanisms of the changes required to forecast the development of eutrophication processes.

Thus, the adaptive approach to the analysis of geoecological data, in contrast to the regressive approach based on the hypothesis of the sustainability of identified system linkages, allows for adjusting the generated model both for *characteristics of the modelled system* and for *specific objectives* of environmental management.

References

Bukanova TV, Stont JI (2010) Use of satellite data in the ecological monitoring of the South-Eastern Baltic. In: Proceedings of the 2nd Baltic green belt forum. Towards sustainable development of the Baltic Sea coast, pp 235–239

Bukanova T, Stont Z, Goushchin O (2013) Sea surface temperature trends in the South–Eastern Baltic from satellite data. In: Proceedings of BSSC 9th Baltic Sea Science Congress: new horizons for Baltic Sea science. Coastal Research and Planning Institute of Klaipeda University, p 145

Carstensen J et al (2013) Approaches and methods for eutrophication target setting in the Baltic Sea region. In: Baltic Sea environment proceedings, vol 133. Helsinki Commission, Helsinki

Kholling KS (ed) (1981) Ekologicheskiesistemy. Adaptivnayaotsenkaiupravlenie (Ecological systems. Adaptive assessment and management). Moscow, Mir Publ. (in Russ.)

Novikov AM, Novikov DA (2011) Metodologiyanauchnogoissledovaniya (Methodology of scientific research). Librokom, Moscow (in Russian)

Störmer O (2011) Climate change impacts on coastal waters of the Baltic Sea. In: Schernewski G, Hofstede J, Neumann T (eds) Global change and Baltic coastal zones, coastal research library-series, vol 1. Springer, Dordrecht, pp 51–69

Walther GR, Post E, Convey P, Menzel A, Parmesan C, Beebee TJ, Fromentin JM, Hoegh-Goldberg O, Bairlein F (2002) Ecological responses to recent climate change. Nature 416:389–395

The Anthropogenic Impact on the Abrasion Coast of the Kaliningrad Region on the Example of the Village of Donskoe

Nikolay S. Belov⬤ and Aleksandr R. Danchenkov⬤

Abstract

Anthropogenic impact on the coastal zone can be seen across the coastal regions. Most often it manifests itself in the form of tourism and recreational activities, as well as in residential and social infrastructure. The Kaliningrad region happens to experience another kind of impact which is illegal amber extraction. This led to the virtual loss of the potential tourist attraction object, the coastline in the village of Donskoe. The conducted research has shown that the volume of the displaced material due to illegal activities exceeds that of the storm effect. A previously stable site turned into a destroyed one, when an increase in wave-denudation can result in problems in the cottage settlement there. Such illegal activities are specific for the Baltic sea region and have gained additional distribution not only on the western coast, but also on the northern coast of the Sambian Peninsula. The impact of human activity on the process of activating erosion geological processes in certain areas raises concerns for the future state of these territories.

Keywords

Coastal zone · Laser scanning · UAV · Abrasion · Baltic Sea

N. S. Belov (✉) · A. R. Danchenkov
Institute of Environmental Management, Urban Development and Spatial Planning, Immanuel Kant Baltic Federal University, Kaliningrad, Russia
e-mail: belovns@gmail.com

A. R. Danchenkov
Shirshov Institute of Oceanology, Russian Academy of Sciences, Moscow, Russia

© Springer Nature Switzerland AG 2020
G. Fedorov et al. (eds.), *Baltic Region—The Region of Cooperation*,
Springer Proceedings in Earth and Environmental Sciences,
https://doi.org/10.1007/978-3-030-14519-4_19

Introduction

Assessment of the coastal zone condition is a well-sanctioned problem in the scientific literature. One of the first works on the state of the coastal zone of the Kaliningrad region is "Notes on the Baltic Coast from Pillau to the Curonian Spit" by Johann Christian Wutzke (1767–1842), published in the Prussian Provincial Notes journal in 1829–1830 (Ryabkova and Levchenkov 2016). In general, the observation period of the condition dynamics for the region is more than 150 years. In recent years, problems related to the activation of exogenous geological processes have been actively discussed (Burnashev et al. 2008; Burnashov 2011; Bobykina and Karmanov 2014; Boldyrev et al. 2010). At the same time, the dynamics of the Baltic coastal zone is considered not only in the context of classical coastal zone formation processes, but also in the context of global climate changes (Harff et al. 2016, 2017; Strandmark et al. 2015).

Frequently, erosion and fracture processes are considered in the example of dune-ridge complexes (Szmytkiewicz and Zabuski 2017; Bugajny et al. 2015; Pruszak et al. 2008; Szmytkiewicz and Różyński 2016; Ostrowski et al. 2016). At the same time, questions of human influence on the Baltic Sea coast development are rarely considered; but when they are, they are usually considered through contributions to pollution (Andersen et al. 2015). Thus, the issues of human influence on the development of the coastal zone in the Kaliningrad region may be of interest, since usually the process of destruction of the bedrock coast is mainly viewed through the prism of classical approaches (Kostrzewski et al. 2015; Uścinowicz et al. 2017).

Materials and Methods

The total length of the sea coastline of the Kaliningrad region is 147 km, of which 39 km is the abrasion coast, and 108 km is accumulative-eroded coast. Natural factors influencing the intensification of abrasion processes can be presented in several groups:

1. storms, storm surges, sea level, beach width, depth of coastal zone;
2. rocks composition and structure, their occurrence, the nature of strata, the influence of landslide processes;
3. the coast morphology (the cliff height), the direction of modern tectonic movements.

The most common factors of anthropogenic impact can also fall into 4 groups:

1. the decrease of sediment runoff into the sea;
2. sedimentary material removal;
3. weakening shore resilience;
4. active construction, disregarding of regional specifics.

The coastline develops most actively within the storm-and-calm cycle, while the impact on the sea coast rocks is determined by the height of the storm surge and the impact of waves. Also, a direct consequence of storm surges is the high dynamics of the beach zone. A combination of longshore currents with a material deficit, which is typical for the Kaliningrad region, leads to predominance of negative trends throughout the coast.

The western coast of the Sambian peninsula witnesses active lithodynamics with two opposing processes. On the one hand, pulp discharge from the amber factory provides the material for industrial accumulation. On the other hand, active land-slide and scree processes are observed in many coastal areas. In addition there is intense construction activity, which in turn can cause the amplification of *exogenous geological processes* (EGP).

In the survey area in the village Donskoe, illegal amber extraction led to the intensification of EGP (Fig. 1). The main constituent rock of the shore ledge is clay sediment, dark gray loam with sand lenses, and interlacing sands of different granularity.

The area is characterized by steep and high slopes with active erosion processes without vegetation cover. Nearby there are two wash basins, which are used as a dump for building debris from the nearby houses (Fig. 2). In 2016, numerous pits made by "black diggers" were found and tracks of small construction equipment were present on the site with active erosion processes, which indicates illegal activities in this sector. In 2017 there was also evidence of illegal amber extraction.

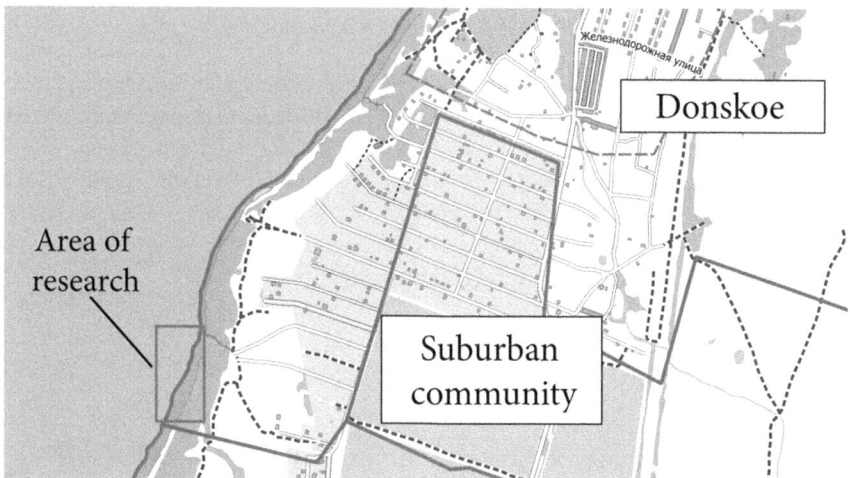

Fig. 1 Survey area

Fig. 2 Areas of most active erosion processes

As it can be seen, the nearest buildings are not far away (maximum 71 m from the site with EGP) and, given the increased erosion processes, may threaten the capital structures.

The work was carried out in 2016–2017. A Topcon GLS 1500 laser geodetic scanner and a DJI PHANTOM 4 were used. The measurements were taken during the summer period to balance out the possible impact of wind and waves. Scanner location points were coordinated using the satellite GPS of the geodetic class Topcon GR5 in the RTK mode. The accuracy of the coordinates was 2 cm in height and 2 cm in plan. The flight mission of DJI Phantom was set by means of Pix4d, coordination was carried out using ground control points (GCP).

Results and Discussion

The section on which the EGP is most actively manifested was limited to 90 m × 50 m in the framework of calculations. The volumetric indicators of the coastal zone transformation from 2016 to 2017 are presented in Tables 1 and 2.

As can be seen, the dynamics of the studied area is insignificant, which is due primarily to the absence of major storm events and a moderate amount of precipitation in 2016–2017 (Fig. 3). Western wind prevailed, which is a characteristic

Table 1 Indicators of the coastal zone dynamics in the survey area

	2016	2017
The maximum width of the beach (m)	11	14
The minimal width of the beach (m)	5	6
The undercliff height (m)	31	31
Site area (m^2)	3624.52	3816.18
Material volume (m^3)	31,577.46	30,972.26

Table 2 The difference in volumes between 2016 and 2017

Volume of the material moved (m^3)	2745.83
The volume of the material moved within the site (m^3)	2140.63
Volume of destruction (m^3)	819.73
Volume of accumulation (m^3)	1320.90
Volume of material lost irretrievably (m^3)	605.2

feature of the South-Eastern Baltic. The wind speed was mostly between 5 and 12 km/h. Major storm activity over the period of fall 2016—spring 2017 was not recorded, which is not typical for the region.

Such subtle change in the area is associated primarily with the movement of material on the site and is hardly the effect of the anthropogenic activity. But the volume values do lead us to some interesting conclusions. The total balance in the studied area is characterized by a negative value of 605.2 m^3. This figure directly confirms low storm impact, as in the Kaliningrad region the maximum impact to EGP development is linked to the storm activity, which gets higher during the cold season (Bobykina and Stont 2015).

Comparative analysis of relief profiles from the years 2016 to 2017 shows that significant changes in the coastal zone relief are observed in the middle part of the shore ledge. The accumulation of the displaced material is located in the lower part of the slope in the range of 4–8 m. The beach was not affected (Fig. 4).

At the same time, the analysis of the movement of volumes within the studied area (Fig. 5) shows no wave erosion. Virtually the entire volume of displaced material remains in the coastal zone. There are no coastal niches. At the same time, the area of accumulation is located at the foot of the coastal ledge, and is absent within the beach.

Analysis of the reasons for the increasing gravitational slope processes led us to the following conclusions. Their main cause was the illegal amber extraction in the lower part of the slope, which was recorded between 2012 and 2016 (Fig. 6). In 2015, due to illegal excavations, there was a local collapse in the volume of 400 km^3, as a result of which one vacationer was killed.

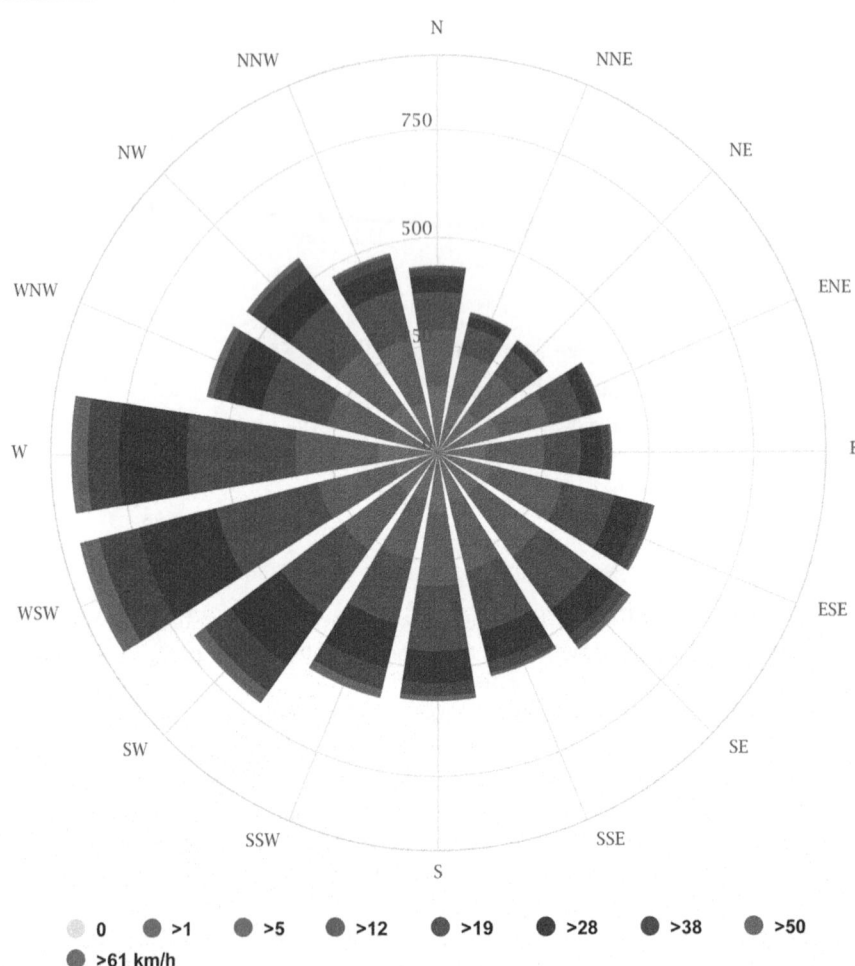

Fig. 3 Wind rose for the period of 2016–2017

The volume of excavations is impressive, because until 2016 they were carried out exclusively by hand. At the foot of the slope there were 3 pits about 4–5 m deep and 4 pits 2 m deep, the average amount of excavated material from one pit was 30 m³ or 48 tons. The excavated material was stockpiled at the sea side in an attempt to create a sort of protection against storms. Thus, illegal activities led to formation of a man-made wave-cut niche.

During 2016, we documented the moments of using light construction equipment for the purpose of amber extraction. This was the main cause of the collapse in the period of fall 2016—spring 2017. The starting point was, apparently, naval exercises and acoustic impact from the shots. A positive development was the

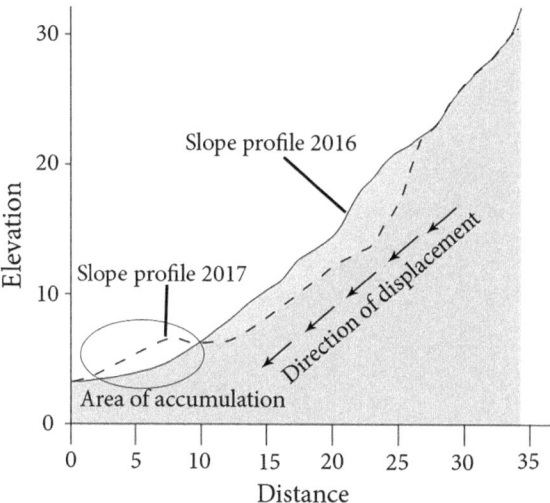

Fig. 4 Slope profiles in 2016 and 2017

Fig. 5 UAV data in 2017

almost complete absence of storm activity, resulting in the absence of wave erosion, which in turn makes illegal activity on this site virtually impossible, since its current condition requires the use of heavy construction equipment. Thus, the contribution of illegal activities that led to the collapse of the coastal slope is easily noticeable.

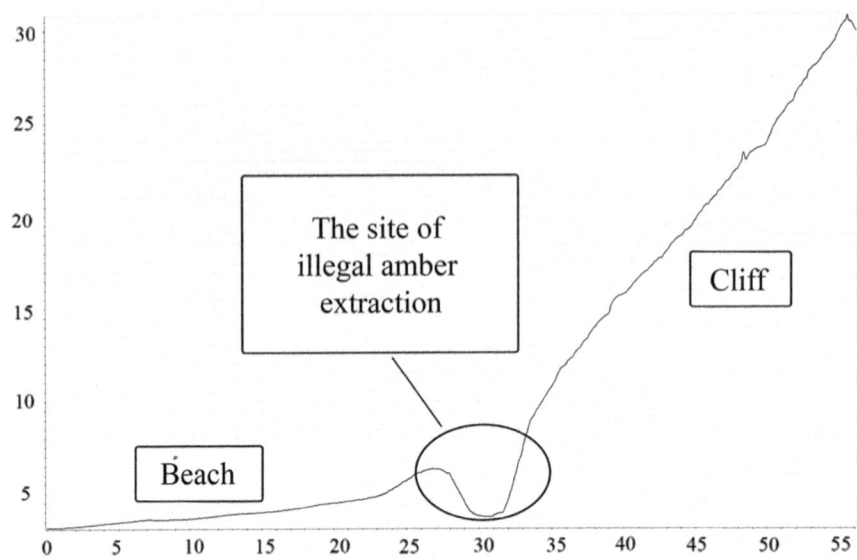

Fig. 6 Slope profile in 2014 with an example of illegal amber extraction

At the same time, the impact of human activity in this area on the process of collapse exceeds that of natural processes. It would take at least 4–5 storm seasons to achieve a similar effect without human interference.

Discussion and Prognoses

The assessment of human impact on erosion processes on the coast is usually viewed through the lens of climate change (Semeoshenkova and Newton 2015), through the impact of large marine infrastructure objects in relation to storm activities (de Boer et al. 2017), or through the interaction of infrastructure with the coastal zone as a whole (Semeoshenkova et al. 2017; Pereira et al. 2016). It is obvious that such locally concentrated phenomena of anthropogenic activity as the ones considered in this study do not attract wide attention. In case of the Kaliningrad region, this issue becomes more and more explored, since there is a negative situation associated with acute material shortages. A preliminary estimation of the material shortage makes 40,000,000 m^3 (Blazhchishin and Boldyrev 1999). In such a situation, any manifestations that led to the intensification of erosion processes should be analyzed through compensation of the caused damage. Adverse effects associated with illegal production of amber within the coastal zone, have now spread from the West coast to the North. Particular difficulty is attached to the fact that the restoration of these territories requires much more money than can be obtained under the current legislation by imposing penalties on individuals who conduct illegal activities. In addition, due to the legislation tightening, a number of individuals got engaged in illegal amber extraction in the coastal zone where depth does not exceed 10 m, which may lead to a change in certain areas along the coastal currents, and, in turn, may affect the formation of the coastal zone.

Acknowledgements The work was supported by IKBFU «5-100» Russian academic excellence project. Data analysis and interpretation was done with a support of the state assignment of IO RAS (Theme No. 0149-2019-0013).

References

Andersen JH, Halpern BS, Korpinen S, Murray C, Reker J (2015) Baltic Sea biodiversity status vs. cumulative human pressures. Estuar Coast Shelf Sci 161:88–92. https://doi.org/10.1038/ncomms8615

Blazhchishin AI, Boldyrev VL (1999) Resursy inertnykh materialov na Kaliningradskom poberezh'e (Resources of inert materials on the Kaliningrad coast). In: EkologicheskieproblemyKaliningradskoioblastii Yugo-VostochnoiBaltiki. Kaliningrad, pp 76–79 (in Russ.)

Bobykina VP, Karmanov KV (2014) K geoekologii beregov Kaliningradskoi oblasti (po rezul'tatam monitoringa) (On the geoecology of the shores of the Kaliningrad region (based on monitoring results). Izvestiya KGTU 35:44–56 (in Russ.)

Bobykina VP, Stont Zh I (2015) O zimnei shtormovoi aktivnosti 2011–2012 gg. i ee posledstviyakh dlya poberezh'' yayugo-vostochnoi Baltiki (About the winter storm activity 2011–2012 and its consequences for the south-east Baltic). Vodnyeresursy 42(3):322–322 (in Russ.)

Boer de W, Huisman B, Yoo J, McCall R, Scheel F, Swinkels C, Friedman J, Deigaard R (2017) Understanding coastal erosion processes at the Korean east coast. In: Proceedings of the conference coastal dynamics 2017, Helsingør, Denmark, pp 1336–1347

Boldyrev VL, Bobykina VP, Chubarenko BV, Burnashev EM, Karmanov KV (2010) Abrazionnye protsessy na beregakh Yugo-Vostochnoi Baltiki (Abrasion processes on the shores of the South-Eastern Baltic). In: Proceedings of the XXIII international coastal conference in honor of the centenary of the birth of V.P. Zenkovich, St. Petersburg, 5–9 October 2013, pp 54–55 (in Russ.)

Bugajny N, Furmańczyk K, Dudzińska-Nowak J (2015) Application of XBeach to model a storm response on a sandy spit at the southern Baltic. Oceanol Hydrobiol Stud 44(4):552–562. https://doi.org/10.1515/ohs-2015-0052

Burnashev EM, Boldyrev VL, Bobykina VP (2008) Fotomonitoring kak naglyadnyi pokazatel' dinamiki pribrezhnoi zony na primere beregov Kaliningradskoi oblasti (Photomonitoring as a visual indicator of the dynamics of the coastal zone by the example of the shores of the Kaliningrad Region). In: Proceedings of the International Conference, Kaliningrad, pp 39–41 (in Russ.)

Burnashov EM (2011) Sovremennaya dinamika morskogo poberezh'ya Kaliningradskoi oblasti po dannym ezhegodnykh monitoringovykh issledovanii. Voprosy sovremennoi nauki i praktiki (Modern dynamics of the coast of the Kaliningrad region according to the annual monitoring studies). University of VI Vernadsky, vol 2, pp 10–17 (in Russ.)

Harff J, Jöns H, Rosentau A (2016) Interrelation of geosphere, climate processes and anthroposphere in the Baltic Sea basin during the Holocene. In: Multiple drivers for earth system changes in the Baltic Sea region, vol 3. https://doi.org/10.2112/04-0217.1

Harff J, Furmańczyk K, Von Storch H (eds) (2017) Coastline changes of the Baltic Sea from South to East: past and future projection, vol 19. Springer, Berlin. https://doi.org/10.1007/978-3-319-49894-2

Kostrzewski A, Zwoliński Z, Winowski M, Tylkowski J, Samołyk M (2015) Cliff top recession rate and cliff hazards for the sea coast of Wolin Island (Southern Baltic). Baltica 28(2). https:// doi.org/10.5200/baltica.2015.28.10

Ostrowski R, Schönhofer J, Szmytkiewicz P (2016) South Baltic representative coastal field surveys, including monitoring at the Coastal Research Station in Lubiatowo, Poland. J Mar Syst 162:89–97

Pereira LC, Trindade WN, da Silva IR, Vila-Concejo A, Short AD (2016) Maranhão beach systems, including the human impact on São Luís Beaches. In: Brazilian Beach systems. Springer, Cham, pp 125–152. https://doi.org/10.1007/978-3-319-30394-9

Pruszak Z, Rozynski G, Szmytkiewicz P (2008) Megascale rhythmic shoreline forms on a beach with multiple bars. Oceanologia 50(2):183–203

Ryabkova OI, Levchenkov AV (2016) Izuchenie poberezh'ya Sambiiskogo poluostrova: vklad nemetskikh, sovetskikh i rossiiskikh uchenykh (Study of the coast of the Sambian Peninsula: the contribution of German, Soviet and Russian scientists). Vestnik of IKBFU, SER. Nat Med Sci 3:44–70 (in Russ.)

Semeoshenkova V, Newton A (2015) Overview of erosion and beach quality issues in three Southern European countries: Portugal, Spain and Italy. Ocean Coast Manag 118:12–21. https://doi.org/10.1016/j.ocecoaman.2015.08.013

Semeoshenkova V, Newton A, Rojas M, Piccolo MC, Bustos ML, Cisneros M, Berninsone LG (2017) A combined DPSIR and SAF approach for the adaptive management of beach erosion in Monte Hermoso and Pehuen Co (Argentina). Ocean Coast Manag 143:63–73. https://doi.org/10.1016/j.ocecoaman.2016.04.015

Strandmark A, Bring A, Cousins S, Destouni G, Kautsky H, Kolb G, Torre-Castro M, Hambäck PA (2015) Climate change effects on the Baltic Sea borderland between land and sea. Ambio 44(1):28–38. https://doi.org/10.1007/s13280-014-0586-8

Szmytkiewicz P, Różyński G (2016) Infragravity waves at a dissipative shore with multiple bars: recent evidence. J Waterway, Port, Coastal, Ocean Eng 142(5). https://doi.org/10.2112/04-0293.1

Szmytkiewicz P, Zabuski L (2017) Analysis of Dune erosion on the coast of South Baltic Sea taking into account dune landslide processes. Archiv Hydro-Eng Environ Mech 64(1):3–15. https://doi.org/10.1515/heem-2017-0001

Uścinowicz G, Jurys L, Szarafin T (2017) The development of unconsolidated sedimentary coastal cliffs (PobrzeżeKaszubskie, Northern Poland). Geol Q 61(2):491–501

Economy of the Communities of the "Forest" Neolithic in the South-Eastern Baltic

Alexander A. Strelkovsky⊙ and Gennady V. Kretinin⊙

Abstract

About six and a half thousand years ago the peoples of local Mesolithic cultures began to master pottery production in the territory of the South-Eastern Baltic. However, the economy and technology of making tools had not changed since the Mesolithic and the Upper Paleolithic. Such cultures are the representatives of the 'Forest' Neolithic. One of them is the Zedmar culture that was the leading one of the early Neolithic in the region which bordered on both Western cultures with a producing economy and Eastern cultures with an appropriating economy. This neighborhood affected primarily the economy of the culture in question. The aim of the study is to identify the characteristic features of the economy of the Zedmar culture communities as representatives of the 'Forest' Neolithic. For this purpose, the artifacts from excavations on the sites of this culture carried out by German, Soviet, modern Russian and Polish researchers were examined, and in 2009 the author participated in survey work carried out on the sites of Zedmar A and D. Theoretical and practical studies showed the decisive importance of two factors for the development of specific features of Zedmar culture economy: namely, the natural one—in the formation of a complex-adapting economy with a large number of Mesolithic features, and the cultural one—in the presence of underdeveloped agriculture and cattle breeding. There is also a hypothesis about the presence of a specific type of gathering in the territory of the South-Eastern Baltic.

A. A. Strelkovsky (✉) · G. V. Kretinin
Institute of Environmental Management, Urban Development and Spatial Planning,
Immanuel Kant Baltic Federal University, Kaliningrad, Russia
e-mail: a-a-strelkovskij@yandex.ru

G. V. Kretinin
Institute for the Humanities, Immanuel Kant Baltic Federal University,
Kaliningrad, Russia

© Springer Nature Switzerland AG 2020
G. Fedorov et al. (eds.), *Baltic Region—The Region of Cooperation*,
Springer Proceedings in Earth and Environmental Sciences,
https://doi.org/10.1007/978-3-030-14519-4_20

Keywords
'Forest' neolithic · Zedmar culture · Complex adaptation · Environmentalism

Introduction

In the old historiography devoted to the early Neolithic of Eastern Europe, there was one peculiar flaw observed in one sphere—the Neolithic was understood only as an era of acquiring new technologies of stone processing and the emergence of ceramics and producing economy. It was very difficult to get away from such an understanding of the essence of the Neolithic. When we say "Neolithic", we mean the cultural and historical phase of the human society development, which involves a change of the previous foundations of the society in community ties (both internal and external); a change in people's lifestyle due to the producing economy or more sophisticated tools.

As for the territory of the Baltic region, such a term as the 'Forest' Neolithic used in Russian archaeological literature is the most important here. Actually, throughout the entire forest zone of Eastern Europe, which extends from the Baltic to the Urals and from the White Sea to the Middle Volga and the Oka, the economy of the appropriating type persisted during the Neolithic era. This can be explained by specific natural conditions, which did not stimulate much the transition to new forms of economy. Thus, during the Early Neolithic, the producing forms of economy were not familiar to this territory. As a result, hunting, fishing and gathering remained the basis of economy (Timofeyev et al. 2004). The first traces of agriculture and cattle breeding are observed in the middle and late Neolithic, i.e. at the end of the 4th millennium BC. Agriculture and cattle breeding acquired the dominant role only at the beginning of the 2 millennium BC, when a different kind of culture had already existed in this territory. Proceeding from this, there is a rather large time discrepancy in the natural process of the evolution of economic activity.

As for the very definition of this Neolithic as 'Forest', it is mainly used in Russian literature and means a set of cultures of hunters and gatherers who lived in the forest zone of Eurasia and who were familiar with pottery production (Russian Academy of Sciences 1996).

Polish researchers call this kind of Neolithic the "paraneolithic" (Malinowski 1991; Gumiński 2003); and German, Scandinavian and other researchers generally consider the artifacts of this period and region to be the Mesolithic, since producing economy had not formed yet. In Western literature there is also a definition of the "subneolithic" (Malinowski 1991), meaning the culture of hunters and gatherers with the first signs of the domestication of animals. Ceramics production and the degree of sedentism do not significantly matter for this definition. It rather shows the process of neolithization of the region as uneven and non-linear—now with exposure to, and then with rejection of progressive forms of economy (Russian Academy of Sciences 1996). There is also a definition of the "Ceramic Mesolithic" (Gumiński 1999). Tadeusz Galinski in his lengthy article tried to compare the two definitions ('Forest' Neolithic and Ceramic Mesolithic) and he came to the conclusion that the definitions were based on different criteria. An economic criterion is in the basis of one definition and a technological-typological criterion is applied for the other, and they can even complement each other in some cases (Galiński 1991).

Neolithic cultures of the forest zone of the Eastern Baltic are of particular interest in connection with the idea of the 'Forest' Neolithic. All the innovations of the Neolithic mentioned above have no relation to them. The development of the Neolithic cultures, as we can observe it in Southern Europe, is getting a noticeable chronological extension on the territory of the Baltic region. However, it would be incorrect to say that the representatives of these communities were too backward and that they could change their lifestyle only through the invasion. Its own way of life and its specific economy was formed on the territory of Eastern Europe, which only could exist here during this period, because this territory did not have the resources that were in the South and that led to the spread of producing economy and the change in the foundations of lifestyle. Nor was it necessary for the Neolithic people of the region. Due to the complex economy management methods (Kolcov 2011), albeit non-producing economy, the societies of the 'Forest' Neolithic cultures did not undergo a necessary crisis of the appropriating economy, otherwise they would not have survived on average from the middle of the 4 to the end of the 3 millennium BC.

Among the natural features that did not contribute to the development of agriculture were the heavily forested areas and soils difficult for cultivation and requiring more sophisticated tools.

Thus, the principles of socio-economic adaptation worked out during the Stone Age with the availability of rich food resources contributed to some conservation of both economic activity and social structure. However, it should be noted that the availability of food resources played a role in the transition from nomadic and semi-nomadic to a relatively sedentary life. The displacement of people in the era of the 'Forest' Neolithic was no longer associated with the migration of wild animals or with a decrease in their numbers, but it was to a greater extent connected with the surface water level in inland water bodies (due to flooding) (Levkovskaya 2000).

The Zedmar culture as a representative of the 'Forest' Neolithic is of great interest for science because its artifacts quite clearly show us the connection of this culture with both Western cultures with a producing economy and with Eastern cultures with an appropriating economy. The culture in question was a kind of boundary between the two "worlds" or a sort of buffer zone that absorbed the features from both sides (Timofeev 1987, 1996).

Methods

In the study there were used general scientific methods (analysis and classification, the principle of description and systematization); archaeological (comparative-typological, stratigraphic) as well as the methods of field research that serve as a basis for the first group of methods. The practical part of the work is that the author personally participated in survey work in 2009 and established the existence of a rich cultural layer on the site. This type of study of the site was only preliminary in nature; it was necessary to find a cultural layer in the northern part of the settlement of Zedmar D (2 pits of 1 m^2. were made) and on the island, but on the west and north sides—the least studied areas (4 pits of 1 m^2. were made). Moreover, it was necessary to find out about the existence of a cultural layer in the easternmost part of Zedmar A settlement (2 pits of 1 m^2. were made) and the southernmost part of the settlement of Zedmar D (2 pits of 1 m^2. were made). Thanks to this practical part of the study there was some new data confirming and correcting old information (Figs. 1 and 2; Table 1).

Results

The leading role of hunting in the economy. The main stage in the development of the Zedmar culture fell on the climatic maximum of the Holocene, which of course could not contribute to the transition to a producing economy as the abundance of

Fig. 1 Zedmar swamp (satellite snapshot). Highlighted Zedmar A and D

Fig. 2 Zedmar D, Schurf 1, 2009

animals in forests strengthens the position of hunting in economy. For example, among the bone remains from the settlement of Zedmar D, the following species are represented: wild boar—240/11, red deer—189/10, aurochs—219/9, moose—47/4, roe deer—41/5, wild horse—11/2, bear—10/3 (Timofeev and Chajkina 2001; Russian Academy of Sciences 1996, pp. 164–165). The signs of cattle breeding give only the evidence of "the first phase of the agricultural frontier development" according to M. Zvelebil and P. Rowley-Conwy, which is characterized by the relations of a kind of "cooperation" between communities of different types of economy (Timofeev 1998, p. 275; Antanaitis-Jacobs and Girininkas 2002) and/or intertribal marriage relations. The domesticated animals did not exceed 5% in the diet of the local population and could not affect the economy of the Zedmar culture in any case. The osteological findings of domesticated animals are located on the sites of the Zedmar culture in the form of several bones of a couple of sheep, goats and cattle. Meanwhile, the bones of wild animals are found many times more often.

Fishing took the second or third place in the economic activities of the Neolithic people who inhabited the Zedmar sites. This can be seen in the ratio of fish remnants and the bones of animals and birds. Thus, in the cultural layer of the settlement of Zedmar D, about 5300 fragments of the bones of animals and birds and only 60 fragments of fish bones were found. The complete absence of weights for fishing nets and remnants of floats and other evidence in favor of net fishing—all

Table 1 Spore-pollen diagram of sediments of the multi-layered settlement of Zedmar A with the cultural layer of the Atlantic stage of settlement (associated with its top) with the earliest pollen findings of cultivated cereal

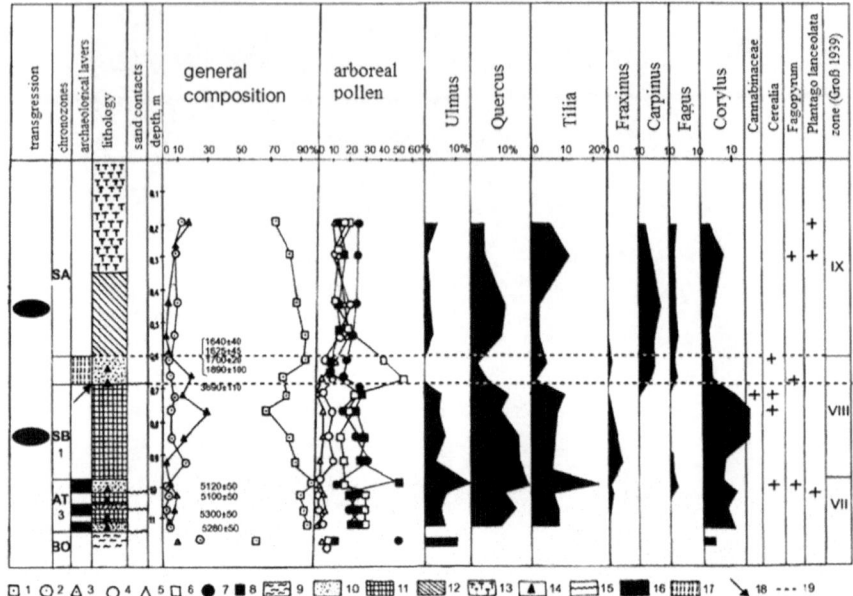

Source Levkovskaya and Timoveef (2004)

Legend 1—amount of pollen from tree species; 2—amount of pollen from herb-shrub plants; 3—spores; 4–7—pollen: birch, 5—spruce, 6—alder, 7—pine; 8—the sum of pollen from broad-leaved trees; 9—silt; 10—sand; 11–12—various in coloring of githia (sapropels); 13—peat; 14—sampling points for radiocarbon dating; 15—levels of sand contacts; 16—cultural layer of the Atlantic stage of colonization and finding of pollen from the first cereals; 17—cultural layer of Early Iron time; 18—level of the roof of Githia with the findings of cultivated cereals of the times of the CWC; 19—levels of breakthroughs in sedimentation

these indicate the underdevelopment of wide-scale fishing and the non-random ratio of bones (Timofeev 1991). At the site of Shchepanka 8, 8% of more than 5400 bones were identified as fish remains.

Such a situation with fishing in the system of economy of the Zedmar Neolithic population is peculiar and non-typical of the Neolithic sites in the Eastern Baltic. For example, a fish weir was found at the Zvidze settlement (Loze 1986), and over 2367 definable fish bones were found in such settlements in Eastern Latvia as Lagaza, Abora I and Eini (Loze 1979). I. Loze notes that "a cluster of fish scales and vertebrae in the form of a continuous layer over a large area" was discovered near the fireplaces (Loze 1979).

The features of gathering and agriculture. A. A. Sejbutis gives an example of the well-known phenomenon of Holocene deposits in North-Western Europe that is typical of the Kaliningrad region—the boreal maximum of the hazel pollen.

The area of abundant growth of hazel groves is pine forests which in the Mesolithic and Neolithic period occupied almost the entire region. Some authoritative researchers, F. Firbas and M. I. Neishtadt among them, argue that during the Mesolithic and the Neolithic there were pure hazel woods that were not covered with a layer of tree species (Sejbutis 1982). Such wide distribution of hazel and its practical separation from other species of trees can be explained in two ways: either due to a change in physical and geographical conditions or by the anthropogenic factor (Sejbutis 1982). The second option seems to be a little more preferable when viewed in connection with the English doctrine of environmentalism, and specifically with Eric Higgs (Kleyn 2011, p. 356) who spoke about the presence of random occurrences of some late natural processes at an earlier time (Sejbutis 1982), namely deliberate burning (elimination) of the forest for a more abundant harvest of nuts that occurred during the Mesolithic period. The hazel is known to quickly spread over an open territory even without the help of people. The second option is supported by an unknown stimulating factor which is manifested in the rise of the hazel curve in comparison with broad-leaved species which, by the way, disappeared in the late Neolithic when agriculture had already spread.

It is possible that people of the early Neolithic age, living in a heavily forested area or a wooded area near water bodies, somehow contributed to the elimination of trees for further allocation of this cleared area for the hazel which was used as food.

Thus, at the end of the Mesolithic on the territory of the South-Eastern Baltic region there might have appeared a local kind of gathering with some signs of producing. The abundance of tools for processing wood found at the site of Utinoye Boloto I (Duck Swamp I) and having the Mesolithic origin were very much suitable for cutting trees (Timofeev and Chajkina 2001).

According to the stratigraphy of the Zedmar sites (Zedmar A and D, Utinoye Boloto I, Dudka I, Szczepanki 8) and the two sets of radiocarbon dates, these places were populated twice: the early stage of 4430-3640 cal BC, with the pollen of cultivated plants (no more than 8–10%) and the basic ceramic material of the Zedmar type; the later stage of 3100–2880 cal BC, with the pollen of cultivated plants and the main ceramic material, such as the Corded Ware and the ceramics of the Globular Amphora Culture (Russian Academy of Sciences 1996, p. 164; Timofeyev et al. 2004, pp. 40–41).

The pollen of cultivated plants (*cerelia*) is recorded in both the non-differentiated and differentiated layers of Zedmar A and D. It was first discovered by H. Gross in wells 1, 4, and 12, and later G.M. Levkovskaya confirmed its presence in the section of the excavation of Zedmar A (Gross 1939, pp. 131, 157; Levkovskaya and Timoveef 2004).

The pollen of cerelia is understood as cultivated cereals. As a rule, when growing it, there are special ruderal and segetal plants—and these were found on Zedmar A together with the pollen from cerelia, fagopyrum (buckwheat) and planta golanceolata (plantain).

All the three types of pollen that indicate agricultural activities are found at three planigraphic levels: early level (end of the Atlantic period, transition from early to middle Neolithic); middle level (end of the Subboreal period, late Neolithic); late level (the beginning of the sub-Atlantic period, the Iron Age) (Levkovskaya and Timoveef 2004).

The cerelia pollen from the cultural layer of the early level, the end of the Atlantic period, coincides in time with the regression of the lake and the culmination of the curve of oak and other broad-leaved species (linden, elm, beech) while the curve of the pine and hazel pollen is falling. The early level of the cerelia pollen deposits corresponds to radiocarbon dating: 3980–3800 cal BC (Bln-2165), 3970–3800 cal BC (Le-1389), 3800–3640 cal BC (Le-1388) (Levkovskaya and Timoveef 2004).

The bones of domestic animals (cattle, sheep, goats) found in very small quantities date back to the same period (Levkovskaya and Timoveef 2004). Probably, domesticated animals appeared as a result of an exchange with other cultures or with the arrival of certain representatives of other cultures onto this territory. This assumption is confirmed by a very small number of bones of domestic animals. However, the population of the Zedmar sites hardly exceeded 40–50 people, so a large number of cattle were not needed.

The hoe-like tools with presumably the traces of loosening the soil are also of great interest. Analogies to this kind of tools can be found both in the Funnel Beaker culture and in the Bug-Dniester culture, but the latter was located at a greater distance and did not directly border on the Zedmar culture.

Much speaks in favor of the fact that the signs of the producing economy that we have cited appeared in the Zedmar culture through the contacts with the Lengyel culture and the Funnel Beaker culture.

Discussion

The following development of economic activity on the Zedmar sites is likely to have happened. Taking into account the maximum level of the hazel pollen in the Mesolithic (5300 ± 50 BP) (Levkovskaya and Timoveef 2004) and just before the appearance of the cerelia pollen (5120 ± 50 BP) (Levkovskaya and Timoveef 2004), the Zedmar culture people were engaged in clearing the light forest surrounding their settlements in order to free the territory and create favorable conditions for the growth of hazel, the nuts of which were widely used for food. Since such activities had a transformational economic character, we can make a reasonable conclusion that they had some producing characteristics.

It is unlikely, but since the area cleared of trees would have an ash-fertilized surface, another place (apart from the floodplain) for sowing cultivated plants appears. It turns out that from the end of the Mesolithic people created conditions and a technological basis for the appearance of slash agriculture on this territory, which would be impossible without the presence of cereal seeds. The seeds of cultivated plants themselves were the result of intercultural contacts as well as domesticated animals only at the beginning of the Middle Neolithic in the region.

When the seeds of cultivated plants appeared, the Zedmar culture communities could well cultivate them. According to V. I. Timofeeva and G. M. Levkovskaya, agriculture could be of floodplain character because the soil in the floodplain was very easy for loosening and planting seeds (Levkovskaya and Timoveef 2004).

But still, agriculture could not play a leading role in the economy of the Zedmar culture communities for several reasons. Firstly, there was no need for it. After all, it is known that agriculture begins to assume its importance only when there is a shortage of food resources gained by hunting or gathering. Secondly, it is very important to emphasize that the Zedmar culture people did not have a large area for sowing and were limited mainly by the area near the floodplains and floodplain territory that appeared due to the lake regression. But since the periods of short-term and wide transgressions replaced the periods of regression quite often, the territory of the sowing was flooded often and for a long time.

So, what are the facts that confirm the above hypothesis? On the one hand, they have a natural scientific origin, but some of them are not attributed directly to the Zedmar geoarchaeological territory. The following facts tell us about the connection between the peculiar type of gathering and the emergence of agriculture:

- The presence of soil erosion due to the influence of fire.
- Two maximums of the hazel pollen are observed immediately before the first appearance of the cerelia pollen, after which the amount of the hazel pollen drops.
- The presence of many tools for woodwork (Utinoye Boloto I) and the tools of hoe-like type with the traces of ground loosening.

The undoubted advantage of this view is that it is at least somehow able to explain the appearance of slash agriculture without breaking the connection with the previous economic practice of people, gathering in this case. However, only further paleobotanical and archaeological data can clarify this issue.

Conclusion

At the late stage of the existence of the Zedmar culture, a stable system of balanced hunting- fishing-gathering and partly agricultural economy developed as an adaptation mechanism, which made it possible for the people to exist in the natural environment of the South-Eastern Baltic region. The economic activities of the Zedmar culture people are different form all other Neolithic cultures of both the Eastern Baltic and the Western part of the Baltic. The main peculiarities of the economy were the following:

- The minor role of fishing in the system of economy (Timofeev 1991), in contrast to other settlements and sites of this period in the Baltic region where seafood in the diet accounted for up to 80% (Dolukhanov 2000);

- The possible specific character of nut gathering with some producing signs (Sejbutis 1982), which was manifested in the purposeful burning of the forest in order to form hazel undergrowth. As is known, hazel quickly occupies the cleared territory without the help of people. An unknown stimulating factor of the hazel curve rise in comparison with broad-leaved species speaks in favor of the second idea;
- The development of the technology of nut gathering for agricultural needs—the (possible) emergence of slash agriculture from the previous specific gathering technology.

It seems that the availability of such extensive faunistic data as well as the individual fishing gear may indicate the fairly well-preserved Paleolithic-Mesolithic traditions in the economy (Gumiński 2001). This is also confirmed by the technology of making tools, which is a slightly more developed Mesolithic version.

It is quite possible that agriculture and cattle breeding in the Zedmar culture would eventually have developed into the leading branch of economy. At a minimum, the prerequisites for this were already created in the 'Forest' Neolithic period (the land cleared from the forest, the presence of cereal seeds). However, the widespread advancement of the tribes of the Corded Ware culture with a producing type of economy prevented this relatively independent economic development of the Zedmar culture.

References

Antanaitis-Jacobs I, Girininkas A (2002) Periodization and chronology of the Neolithic in Lithuania. Archaeologia Baltica 5:9–39

Dolukhanov PM (2000) Istoki Etnosa (Origins of ethnicity). The European House, St. Petersburg (in Russ.)

Galiński T (1991) Uwagi na temat mezolitu ceramicznego I neolitu strefy lesnej na nizu polskim. In: Archeologia Polski, vol XXXVI. Warsaw, pp 5–71

Gross H (1939) Moorgeologische Untersuchung der vorgeschichten Dörfen Zedmar-Bruch. In: Prussia. Zeitschrift für Heimatkunde. Königsberg, pp 101–161

Gumiński W (1999) Kultura Zedmar a kultura Narva. Razem czy osobno. In: Światowit, vol XLII. Warsaw, pp 59–69

Gumiński W (2001) Kultura Zedmar. Na rubieży neolitu «zachodniego». In: Od neolityzacji do początków epoki brązu. Poznan, pp 133–151

Gumiński W (2003) Szczepanki 8. Nowe stanowisko torfowe kultury Zedmar na Mazurach. In: Światowit, vol XLVI, Warszaw, p 53

Kleyn LS (2011) Istoriya arkheologicheskoy mysli (History of archaeological thought), vol I. St. Petersburg (in Russ.)

Kolcov LV (2011) Varianti adaptacii v mezolite lesnoy Evropi (Variants of adaptation in the Mesolithic Forest Europe). In: Paleolit i mezolit Vostochnoy Evropi. IA RAN-Press, Moscow, pp 418–431 (in Russ.)

Levkovskaya GM (2000) Khronologiya i paleografiya ozernykh transgressiy i regressiy epokh mezolita, neolita i ranney bronzy v Vostochnoy Pribaltike (lubanskaya i tsedmarskaya niziny) (Chronology and paleography of lake transgressions and regressions of the Mesolithic, Neolithic and Early Bronze epochs in the Eastern Baltic (Lubana and Zedmar lowlands).

In: Khronologiya neolita Vostochnoy Evropy. Tezisy dokladov mezhdunarodnoy konferentsii posvyashchennoy pamyati N.N. Gurinoy: (The Neolithic Chronology of Eastern Europe. Abstracts of International Conference dedicated to the memory of NN Gurina). St. Petersburg, pp 40–45 (in Russ.)

Levkovskaya GM, Timoveef VI (2004) K hronologii i ekologii nachala zemledeliya v Vostochnoj Pribaltike (To the chronology and ecology of the beginning of agriculture in the Eastern Baltic region). In: Problemi hronologii i etnokulturnih vzaimodejstvij v neolite Evrazii. IIMK RAN-Press, St. Petersburg, pp 88–105 (in Russ.)

Loze IA (1979) Pozdniy neolit i rannyaya bronza Lubanskoy ravniny (Late Neolithic and Early Bronze Lubāns plains). Riga (in Russ.)

Loze IA (1986) Rybolovnyy zakol epokhi neolita na poselenii Zvidze (Fish weirs on the Neolithic settlement Zvidze). KSIA 185:78–82 (in Russ.)

Malinowski T (1991) Sprawa termiligiczna: kultura narewska. In: Archeologia Polski, vol XXXVI. Warsaw, pp 264–267

Russian Academy of Sciences (1996) Neolit Severnoj Evrazii (Neolithic of Northern Eurasia) Nauka, Moscow (in Russ.)

Sejbutis AA (1982) Vazhnejshie cherti razvitiya golocenovoj rastitelnosti v Pribaltike (The most important features of the development of Holocene vegetation in the Baltics). In: Razvitie prirodi territorii SSSR v pozdnem plejstocene i golocene. Nauka, Moscow, pp 162–168 (in Russ.)

Timofeev VI (1987) K probleme svyazej v neolite Baltijskogo regiona (To the problem of interactions in the Neolithic of the Baltic region). In: Proceedings of the Zadachi sovetskoy arheologii v svete reshenij XXVII Sjezda KPSS. Nauka, Moscow, pp 249–250 (in Russ.)

Timofeev VI (1991) Nekotorie dannie o rybolovstve v kamennov meke (Some data on fishing in the Stone Age). In: Rybolovstvo i morskoj promysel v epohu mezolita – rannego metalla v lesnoj i lesostepnoj zone Vostochnoj Evropi. St. Petersburg, pp 87–90 (in Russ.)

Timofeev VI (1996) K probleme Yugo-Vostochnyh svyazej neolita Vostochnoj Pribaltiki (To the issue of the South-Eastern interactions of the Neolithic of the Eastern Baltic States). In: Tverskoj Arheologicheskij sbornik. Tver, pp 183–189 (in Russ.)

Timofeev VI (1998) Zedmarskaya kultura v neolite Vostochnoj Pribaltiki. In: Tverskoj Arheologicheskij sbornik. Tver, pp 273–279 (in Russ.)

Timofeev VI, Chajkina LG (2001) O strukture neoliticheskogo torfyannikovogo poseleniya Utinoe Boloto 1 b Kaliningradskoj oblasti (Vostochnoe poselenie) (On the structure of the Neolithic peat settlement in Utinoye Boloto 1 in the Kaliningrad region (eastern settlement). In: Kamenniy vek evropeiskih ravnin. Sergiev Posad. Russia, pp 211–217 (in Russ.)

Timofeyev VI, Zaytseva GI, Dolukhanov PM, Shukurov AM (2004) Radiouglerodnaya hronologia neolita Severnoj Evrazii (Radiocarbon chronology of the Neolithic of Northern Eurasia). St. Petersburg (in Russ.)

Contemporary Applied Geographic Studies in the Baltic Sea Region: Public Geography

Polish-Russian Small Border Traffic—A Summary Attempt

Renata Anisiewicz

Abstract

The article gives an overview of the conditions and aspects of small border traffic between Poland and the Kaliningrad Oblast, Russian Federation, introduced in 2012 and suspended in 2016. Its impact on the social and economic growth of the region is analysed, and the consequences of its suspension are indicated.

Keywords

Small border traffic · Kaliningrad Oblast · Polish-Russian border · Warmian-Mazurian Voivodship · Tri-City

Introduction

The conditions on the border of Poland and the Kaliningrad Oblast, RF, have a major impact on the social and economic development of the Polish-Russian border region. Up to the end of nineteen eighties, the border remained a barely permeable barrier, which hindered contacts of the border region population. The political and economic transformation of the entire region initiated in the early nineteen nineties, triggered diametric changes, also characteristic for the borders of other countries of Central and Eastern Europe. The transformations increased openness of the Polish-Russian border, varied as it was over time under the impact of various political and economic factors. The greatest change dynamics accompanied Poland's accession to the Schengen area in 2007. Its major consequence was the introduction of the small or local border traffic (LBT) in 2012. Its suspension in the

R. Anisiewicz (✉)
Department of Regional Development Geography, University of Gdańsk, Gdańsk, Poland
e-mail: geora@ug.edu.pl

© Springer Nature Switzerland AG 2020   193
G. Fedorov et al. (eds.), *Baltic Region—The Region of Cooperation*,
Springer Proceedings in Earth and Environmental Sciences,
https://doi.org/10.1007/978-3-030-14519-4_21

mid-2016 justifies the attempt to assess the impact on the social and economic situation of the border region of its four-year duration.

The Conditions Accompanying the Introduction of the Small Border Traffic

Poland's accession to the European Union in 2004 necessitated tightening the country's eastern border seen as the future eastern border of the EU. As early as in 2003, citizens of the countries neighbouring with the future EU member states were obliged to obtain visas; in the case of Poland, this requirement applied to the Russians, Belarussians, and Ukrainians. In retaliation, Russia, Belarus, and Ukraine introduced the same requirement for Polish citizens (Anisiewicz and Palmowski 2014). The fact triggered a gradual slowdown of the border traffic between Poland and the Kaliningrad Oblast, which over five years, immediately preceding the accession, had oscillated around 4 million crossings border annually.

Another major step in the functioning of Poland's eastern border came with the broadening of the Schengen area by nine new states, Poland included, which took place on 21 December 2007. The treaty guarantees the citizens of 26 European states freedom of movement enhanced by the abolition of the internal border control. Alongside abolition inside the EU, its external border was established and the visa procedures harmonised, the latter including e.g. the principles of issuing the so-called Schengen visas[1] to the citizens of third states.

The EU authorities presumed that the main and immediate effect of expanding the Schengen area would reduce border traffic, and consequently generate social and economic problems in the border regions. In order to mitigate the anticipated losses, national authorities were given the option of introducing certain relief measures and simplifying border traffic, e.g. in the form of small border traffic.

As anticipated, the introduction of Schengen visas resulted in gradual dwindling of the number of people crossing the Polish-Russian border, down to a mere 1.28 million in 2009. Signs of an increase were recorded in subsequent years. The formal restrictions in the flow of persons impeded and reduced small border trade in the area, which in earlier years mitigated social and economic problems in the region. The interest of Russians in touring Poland also dropped and affected the standing of Polish companies operating in the tourist trade (Anisiewicz and Palmowski 2014).

To mitigate the negative effects of the broadened Schengen area, the representatives of the European Union and the Russian side sat down to negotiations as early as in October 2007. The EU confirmed its willingness to introduce a special border crossing procedure on the Polish and Lithuanian border, in the form of small

---

[1]Up to 1 June 2007, the residents of the Kaliningrad Oblast, similarly as in the case of the citizens of other countries along the eastern border of the EU, had the right to obtain a free entry visa to the EU. Pursuant to the Sochi agreement between the European Community and Russian Federation signed on 25 June 2006, as of 1 June 2007 the Russians have to pay the visa fee of EUR 35. The same fee is mandatory for Poles applying for a Russian visa.

border traffic, covering the area of 30 km off the border. That would not have included Kaliningrad, therefore the Russians insisted on small border traffic covering the area 50 km off the Polish and the Lithuanian borders (Żukowski 2013). Following perturbations caused by Lithuania's negative stance, in April 2010, Poland and Russia placed a postulate with the EU motioning that small border traffic should include the entire Oblast. A year later, the EU Commission and the European Parliament approved the project. The bilateral agreement signed in Moscow on 14 December 2011 was ratified in Warsaw on 4 May 2012 and in Moscow on 16 June 2012. Small border traffic was initiated on 27 July 2012 (Anisiewicz and Palmowski 2014).

The zone of the small border traffic rights extended to include the entire Kaliningrad Oblast, 11 Polish poviats of the Warmian-Mazurian Voivodship, both directly on the border and those neighbouring thereon, plus the towns of Olsztyn and Elbląg and 4 coastal poviats of the Pomeranian voivodship with the cities of Gdańsk, Sopot, and Gdynia.

The key facilitating measure introduced under the small border traffic agreement involved the option for the inhabitants of the Kaliningrad Oblast and the neighbouring Polish poviats, residing there for at least 3 years, to obtain special passes, which entitled the holders to multiple border crossings and remained valid for 2 to 5 years, for the fee of EUR 20. They did not imply the right to take up a job or pursue business in the neighbouring country, but were intended to facilitate social, cultural, family, and other contacts of economic nature between the population of the border regions (Dudzińska and Dyner 2013).

On 4 July 2016, by virtue of the governmental decision of the Republic of Poland, small border traffic with Russia and Ukraine was suspended. The reasons given were as follows: a threat from the neighbouring countries to the major international events planned in Poland for July that year: the NATO summit in Warsaw, and the World Youth Days in Kraków. Small border traffic with Ukraine was restored on 3 August 2016, whereas the traffic with Russia remains suspended to date.

The Functioning of Small Border Traffic

The rising trend in the Polish-Russian border traffic, observed as early as of the year 2008, indicated potential dynamic growth with the introduction of local border traffic. This was exactly the case. The number of people crossing the border rose steadily up to the year 2014 reaching ca. 6.5 million a year. Following the 2015 political events on Crimea and in Eastern Ukraine, which resulted in economic sanctions imposed on Russia by the EU countries and the ensuing drop of the ruble exchange rate, the border traffic slowed down. However, after a certain drop in the early 2015, the traffic began to rise once again reaching the highest level in the summer months of the same year. An evident decrease in border traffic came as the result of the LBT suspension in the early July 2016, reflected by the fact that the number of border crossings in the third quarter of the year slumped to nearly one half of the figure reported in the previous year quarter, and to ca. 4 million over the whole year 2016 (www.strazgraniczna.pl).

For years, the number of Poles and foreigners noted in border traffic were similar and oscillated around 50%. The proportion clearly changed after the suspension of small border traffic in the third quarter of 2016; the number of Poles travelling to the Kaliningrad Oblast evidently diminished. This fact points to the significance of the LBT facilitation measure to the inhabitants of the Polish side of the border region, and a detailed analysis of their share in the border traffic corroborates the observation. After the introduction of small border traffic, interest in crossing the Polish-Russian border under the new regulations continued to grow both among Poles and the Russians, even though the growing trend lasted longer on the Polish side. The number of Russians crossing the border with LBT cards stabilised at about 400 thousand a quarter as early as at the turn of 2013 and 2014, while the number of Poles who availed themselves to the right exceeded 800 thousand in mid-2014 and did not drop below 700 thousand over several following year quarters, subject only to seasonal fluctuations.

Those evidently different numbers representing the inhabitants of the two sides of the border region who exercised their small border traffic rights translated to the share of the LBT cardholders in the entire border traffic. On the Russian side, the share oscillated between 50 and 60% by year quarter, whereas on the Polish side, LBT clearly dominated and in some year quarters exceeded as much as 97% of the entire Polish border traffic.

An analysis of the structure of the whole border traffic volume explains the differences observed in the interest in the Polish-Russian small border traffic. According to the Central Statistical Office, wealthy inhabitants of Kaliningrad prevailed among the Russians who entered Poland, followed by a lower number of Russians from other parts of the exclave. On the Polish side, a vast majority (95%) represents financially struggling individuals originating from the immediate border area, i.e. up to 30 km off the border (Border traffic and expenses... 2016).

The above figures reflect the declared purposes of visits abroad. For years, the dominating purpose of Poles' visits in the Oblast (ca. 97%) was to purchase goods subject to excise duty (prevailingly fuel), the sale of which on the Polish side used to be an additional or even the sole source of income (Wenerski 2014). The travels to Poland made by the Kaliningrad inhabitants were more varied in nature. In the year 2015, shopping expeditions were declared by fewer than 70% people crossing the border, nearly 14% would travel for recreation and leisure, about 9% crossed Poland in transit, and ca. 7% of the surveyed population pointed to other purposes (Border traffic and expenses... 2016). The wealthier residents of Kaliningrad usually travel for rest or leisure and entertainment, and head primarily to the Tri-City where they visit different districts of Gdańsk, Gdynia, and Sopot, participate in major sports and cultural events, and do shopping in shopping malls (Anisiewicz and Palmowski 2014). The less wealthy Russians take advantage of lower prices and the higher quality of food articles, clothes, electronics, furniture, and building materials, as well as services, e.g. medical care or car repairs (Subocz and Sternicka-Kowalska 2015). They tend to buy goods and services in the localities close to the border, though do not omit other parts of the border-neighbouring voivodships (Border traffic and expenses... 2016).

Small Border Traffic Effects

The introduction of the small border traffic contributed to economic revival of the Polish-Russian border region. This was best reflected by increased trade volumes in border areas on both sides of the border (Dudzińska and Dyner 2013; Wenerski and Kaźmierkiewicz 2013). Quoting after the estimates of the Central Statistical Office, the spending abroad in the region quadrupled in the years 2011–2014. In the year 2014, the Poles spent more than PLN 400 million in the Kaliningrad Oblast, as many as 90% of them travelling under the LBT arrangement, and the Russians spent over PLN 800 million in Poland, including ca. 50% LBT holders (Border traffic and expenses... 2016). Higher trade volumes stimulated the setting up of new trade and service businesses, especially in the border belt of the Warmian-Mazurian Voivodship (Subocz and Sternicka-Kowalska 2015). They also contributed to a dynamic development of the tourist trade, particularly in the Tri-City (Anisiewicz and Palmowski 2014; Dudzińska and Dyner 2013).

The impact of small border traffic in the social sphere was no less important. The availability of LBT cards stimulated increased interest in Poland among the inhabitants of Kaliningrad, the less wealthy included; to the latter, a visit abroad meant not only an opportunity to do shopping cheaper, but also to boast about a trip to another country (Wenerski 2014). Gdańsk gained most in terms of image, as its trade and service offer was relatively efficiently adjusted to the needs of the Russian speaking customers. The benefits of the Warmian-Mazurian voivodship were less palpable (Studzińska and Nowicka 2014). Moreover, the introduction of LBT facilitated scientific and cultural cooperation between schools, universities, and non-governmental organisations, stimulated trade contacts (e.g. between the hoteliers or restaurateurs), and enhanced the collaboration of the border services (Dudzińska and Dyner 2013). This all contributed to the process of overcoming the stereotypes of mutual relationships between the two neighbours: Poland and Russia (Studzińska and Nowicka 2014). The increased border traffic, however, did not generate higher crime rate (smuggling, illegal border crossings, theft, traffic incidents), which had raised the greatest apprehension before the introduction of LBT. The higher number of events of this type recorded at the time was proportionately lower than the increase in border traffic itself (Dudzińska and Dyner 2013).

Conclusion

The suspension of small border traffic in mid-2016, which reduced the number of border crossing by ca. 30% compared to the year 2015, affected the economy of the entire region. Lower spending abroad—which according to the Central Statistical

Office fell by almost 40% on the Polish side and nearly 20% on the Russian side—affected the income of local entrepreneurs, especially in trade and tourism. The curtailing of free border traffic flow touched the inhabitants of the Polish border region who used the LBT easements to a much broader extent than the Russians. The less wealthy inhabitants of Kaliningrad also suffered negative effects, as small border traffic had provided an opportunity of reaching more frequently for cheaper and higher quality goods and services. LBT suspension has had a lesser impact on the number of wealthier Russians crossing the border with visas, estimated at 300 thousand (Wenerski 2014). The fact finds confirmation in the recorded growing spending in the first quarter of the year 2017 compared to the same quarter of 2016 (www.stat.gov.pl).

References

Anisiewicz R, Palmowski T (2014) Small border traffic and cross-border tourism between Poland and the Kaliningrad oblast of the Russian Federation. Quaestiones Geographicae 33(2):79–85

Dudzińska K, Dyner AM (2013) Mały ruch graniczny między obwodem kaliningradzkim a Polską – wyzwania, szanse i zagrożenia. Policy Paper 29(77):1–6

Główny Urząd Statystyczny (2017). www.stat.gov.pl. Accessed 13 Sept 2017

Statistical Office in Rzeszów (2016) Border traffic and expenses made by foreigners in Poland and Poles abroad in 2015. https://rzeszow.stat.gov.pl/en/publications/border-areas/border-traffic-and-expenses-made-by-foreigners-in-poland-and-poles-abroad-in-2015,13,2.html. Accessed 15 Nov 2017

Straż Graniczna (2017). www.strazgraniczna.pl. Accessed 12 Sept 2017

Studzińska D, Nowicka K (2014) Mały ruch graniczny z perspektywy władz lokalnych i mieszkańców. Przykład granicy polsko-rosyjskiej. Prace i Studia Geograficzne 54:275–288

Subocz E, Sternicka-Kowalska M (2015) Korzyści i koszty funkcjonowania umowy o małym ruchu granicznym z Rosją w opiniach mieszkańców terenu przygranicznego. Opuscula Sociologica 2:53–65

Wenerski Ł (2014) Mały ruch graniczny pisany cyrylicą. Instytut Spraw Publicznych, Warszawa

Wenerski Ł, Kaźmierkiewicz P (2013) Krajobraz pogranicza. Perspektywy i doświadczenia funkcjonowania małego ruchu granicznego z Obwodem Kaliningradzkim. Instytut Spraw Publicznych, Warszawa

Żukowski A (2013) Mały ruch graniczny między Polską a Rosją – wstępne konkluzje. In: Kotowicz W, Modzelewski WT, Żukowski A (eds) Polska polityka wschodnia a współpraca zagraniczna województwa warmińsko-mazurskiego. Instytut Nauk Politycznych Uniwersytetu Warmińsko-Mazurskiego, Olsztyn, pp 55–63

The Current Situation and the Prospects for Polish Ferry Shipping in the Baltic Sea in the 1st Half of the 21st Century

Marcin Połom⊙

Abstract

The Baltic Sea region is an attractive area for the development of ferry routes. The relatively low distances between seaports in Nordic countries and the south coast of the Baltic makes these routes quick and financially viable. In recent years Polish and foreign shipowners operating out of Gdynia, Gdańsk and Świnoujście have been gaining share in this market. The article outlines the development of Polish passenger ferry services in the Baltic Sea. An attempt has been made to characterise the role of Polish ferry shipowners and Polish-operated ferry routes in passenger shipping on the Baltic Sea market. The structure of current routes and the volume of passenger services have been presented, the prospects for the ferry shipping market have been described.

Keywords

Maritime transport · Ferry shipping · Passenger ferries · The Baltic Sea · Poland

Introduction

The Baltic Sea Region is an undisputed leader in the number of ferry operators and the volume of service. To quote Urbanyi-Popiołek (2012a), the Baltic's position in the global ferry shipping market is very impressive. In recent years there were between 20 and almost 30 ferry shipowners on the Baltic, operating multiple destinations between Nordic countries and Germany, Poland and the Baltic States

M. Połom (✉)
Department of Regional Development Geography, Faculty of Oceanography
and Geography, Institute of Geography, University of Gdańsk, Gdańsk, Poland
e-mail: marcin.polom@ug.edu.pl

© Springer Nature Switzerland AG 2020 199
G. Fedorov et al. (eds.), *Baltic Region—The Region of Cooperation*,
Springer Proceedings in Earth and Environmental Sciences,
https://doi.org/10.1007/978-3-030-14519-4_22

(Estonia, Lithuania and Latvia). A marginal role in this segment was played by ferry shipping to the ports of the Russian Federation. Czermański (2010) notes that the Baltic Sea region shows considerable specificity, differentiating it from other economic areas of the European Union and can thus be considered separately. This is largely due to the relatively short distances between ports, with the journey taking no more than 30 h (the longest trip, from Germany to Finland) and usually between 6 and 10 h. The proximity of the ports allows ferries to compete with low cost airlines.

Ferry services are most concentrated in the West Baltic, mainly around shipping lines between Germany (Lubeck, Travemünde, Sassnitz), Poland (Świnoujście) and Sweden (Ystad, Trelleborg) and between Denmark and Sweden. Approximately 60% of the total volume of traffic in the Baltic took place in this part of the sea. The route network was least dense in Central Baltic, so in the area which includes the Polish coast with the ports in Gdańsk and Gdynia.

There are currently four ferry shipowners in Poland: two companies owned by the Polish State Treasury and two foreign companies. The main ferry port in Poland is Świnoujście with three operators (Polish Baltic Shipping (Polska Żegluga Bałtycka—PŻB), Unity Line and TT-Line) offering services to Trelleborg and Ystad in Sweden. On both these routes there are 11 RoPax and train/car ferries operating. The single largest passenger route is from Gdynia to Karlskrona operated by the Swedish carrier Stena Line, which allocated four ferries there. The smallest passenger volume is recorded on the Gdańsk to Nynäshamn route. The route is operated by Polska Żegluga Bałtycka with one ferry. Seeing the potential of this line, the Gdynia-based shipowner Stena Line launched a competitive service in October 2017, allocating one RoPax ferry.

Polish ports fit in the so-called Motorways of the Sea concept, as they are located along transport corridors leading from the North to the South of Europe. The concept of Motorways of the Sea has first occurred in European terminology in 2001 in the European Commission's White Paper *European Transport Policy for 2010* (Kotowska 2010). The idea of the Motorways of the Sea is closely related to sustainable mobility policy, in which an important role is played by short-haul shipping and Trans-European Transport Networks TEN-T. The Motorways of the Sea fit both in the scope of short sea shipping and the TEN-T networks (Urbanyi-Popiołek 2010). A Motorway of the Sea should include the infrastructure and the route between at least two ports in different EU member states. Poland and Sweden jointly applied for creating two Motorways of the Sea to connect both countries. The first route would be Gdynia—Karlskrona, the second Świnoujście—Ystad. Classifying both routes as Motorways of the Sea is a key element in the development of ferry shipping in Poland.

Outline of the Development of Ferry Routes in Poland at the Turn of the 20th and 21st Century

The period of economic and political transformation had an impact on the development of the ferry market and on the general standing of Polish Baltic Shipping (PŻB). Initially, in the mid-nineties fleet modernisation was started. In 1997 one of the traditional ferries (Pomerania) was rebuilt, and in the same year the fast catamaran Boomerang was launched on the route to Ystad. It later turned out to have been a completely wrong decision, because the ship's hull was rather delicate and did not perform in winter, especially in storm conditions. Consequently, it was only operated from April to October, trying to secure charters for the remainder of the year. In 1998 the ferry Silesia, a twin to Pomerania, was refurbished to a lesser degree. The growing revenue from the increased volume of services, however, did not cover the cost of fleet refurbishment.

A *Recovery Plan for the Polish Baltic Shipping Co* was drafted in 2000 and implemented in 2001 (Polska Żegluga...). The key project tasks were to organize and restructure the company. In the following years, the ferry Rogalin was sold and replaced with the ferry Scandinavia, purchased from the Swedish operator Rederi AB Gotland, operating under the brand Destination Gotland. Unfortunately, even though the ship was more modern than Rogalin it had already had 23 years of service by then. In 2004 an equally old ferry Wawel was purchased, which was placed on the line to Ystad in 2005, replacing Silesia.

In the following years further attempts were made to restructure Polish Baltic Shipping, and an outside investor was sought through privatisation. Early attempts to sell shares in the company were made since autumn 2009. In 2010 advanced negotiations were held with the Danish shipowner DFDS Seaways, which eventually did not result in selling Polish Baltic Shipping. In 2014 another attempt to sell was made. At that time there was an opportunity to consolidate Polish shipowners, as one of the seven entities willing to purchase Polish Shipping Lines was Polsteam (PŻM). However, the Ministry of Treasury did not shortlist this shipowner for further negotiations, limiting the list of potential buyers to four entities.

This turbulent period for Polish Baltic Shipping also saw changes in demand and serviced routes. In 1995 a new route was opened from Świnoujście to Malmö. In subsequent years four routes out of Świnoujście were maintained, including a seasonal route to Bornholm. Two, and seasonally three ferries sailed from Gdańsk to Nynäshamn and Helsinki. Then the number of services was limited. The first route to be closed was to Helsinki followed by Malmö and Copenhagen.

A shipowner younger than Polish Baltic Shipping is Unity Line, a company currently owned by Polsteam, but originally founded by Żegluga Polska (Polsteam's subsidiary), EuroAfrica Linie Żeglugowe (ESL) and Polish Baltic Shipping. The company was incorporated in 1994. Since the start of its operations in 1995, the company serviced the route from Świnoujście to Ystad, and since 2007—from Świnoujście to Trelleborg.

As the volume of services grew, both companies modernised their ferry fleets under the Unity Line brand. On June 1st, 1995 the route Świnoujście-Ystad saw the launch of Polonia, a ferry built specifically for that route (Pacuk 1997). This was also the launch of the new route and the new brand. At this time also two train/car ferries owned by ESL were operated, i.e. Jan Śniadecki and Mikołaj Kopernik. The latter was replaced in 2008 by a higher capacity ferry bought on the secondary market (Kopernik). In subsequent years the following RoPax ferries were purchased: Galileusz (ESL 2006), Gryf (Polsteam 2005), Wolin (Polsteam 2007) and Skania (Polsteam 2008). The latter was built in the same way as Polonia was built back in 1995 and serves as its complement. Both ferries operate a shuttle service and have the largest passenger capacity of all ships managed by Unity Line.

In 2010–2011 the launch of two new RoPax ferries was planned, Piast and Patria, to be built in Szczecin based on Polish original design. The idea was never implemented, because the prospective builder, Stocznia Szczecińska Nowa, announced bankruptcy. Now Polsteam is planning to join the government *Batory Programme* for building new vessels for ferry shipowners.

In addition to the two Polish shipowners, there are also foreign enterprises operating in Poland. The route from Gdynia to Karlskrona is serviced by Stena Line of Sweden, one of the largest ferry shipowners in Europe, and the route from Świnoujście to Trelleborg—by the TT-Line of Germany. The Swedish shipowner has been present on the ferry market since 1962, initially serving routes between Sweden and Denmark. In the following years the company grew rapidly. Stena Line appeared on the Polish market in 1995, when Lion Queen, then owned by Stena's subsidiary Lion Ferry, sailed for the first time to Karlskrona. Fast growth in passenger volume led to the ferry being replaced by the larger Lion Europe (in 1997). In 1998 Stena Line gained full control over the route Gdynia-Karlskrona and maintained service under the main brand, using the same ferry renamed Stena Europe. In 2001 a second ferry, Stena Traveller, was launched on that route. In 2002 Stena Europe was replaced by Stena Baltica, and in 2004 Stena Traveller was replaced by the larger Stena Nordica. In the following years the line had large growth of passenger and RoRo cargo volumes, leading in 2007 to the introduction of a third ferry, Finnarrow, which was operated until 2010. Then, the even larger Stena Vision was introduced, and in 2011 it was joined by the twin ferry Stena Spirit. In the following years Stena Baltica was replaced by a ferry of the same name but of larger cargo capacity. In August 2017 a fourth ferry, Gute, was introduced.

Since 2014 the route from Świnoujście to Trelleborg is serviced by the German shipowner TT-Line with one ferry Nils Dacke, competing with Polish operators Polish Baltic Shipping and Unity Line.

The Structure of the Polish-Operated Ferry Routes in the Context of the Baltic Sea Market

Ferry operators on the Polish market focussed on destinations in Sweden, connecting them to three Polish ports (Gdańsk, Gdynia and Świnoujście). Both Polish (Polferries, Unity Line) and foreign (Stena Line, TT-Line) shipowners offer exclusively routes to Sweden, especially to the seaports on the south coast: Karlskrona, Ystad and Trelleborg. The only route out of Gdańsk port connects it to the Stockholm conurbation (Nynäshamn). In the past there were many routes to Finland, Denmark, Germany and the United Kingdom (Fig. 1).

Contemporary economic conditions and the fact that profits are mainly generated by shipping cargo, determine the route structure in Central Baltic. The passenger traffic is of less importance, even though it constantly grows in volume. There is noticeable untapped tourist potential, even though the routes from Poland to South

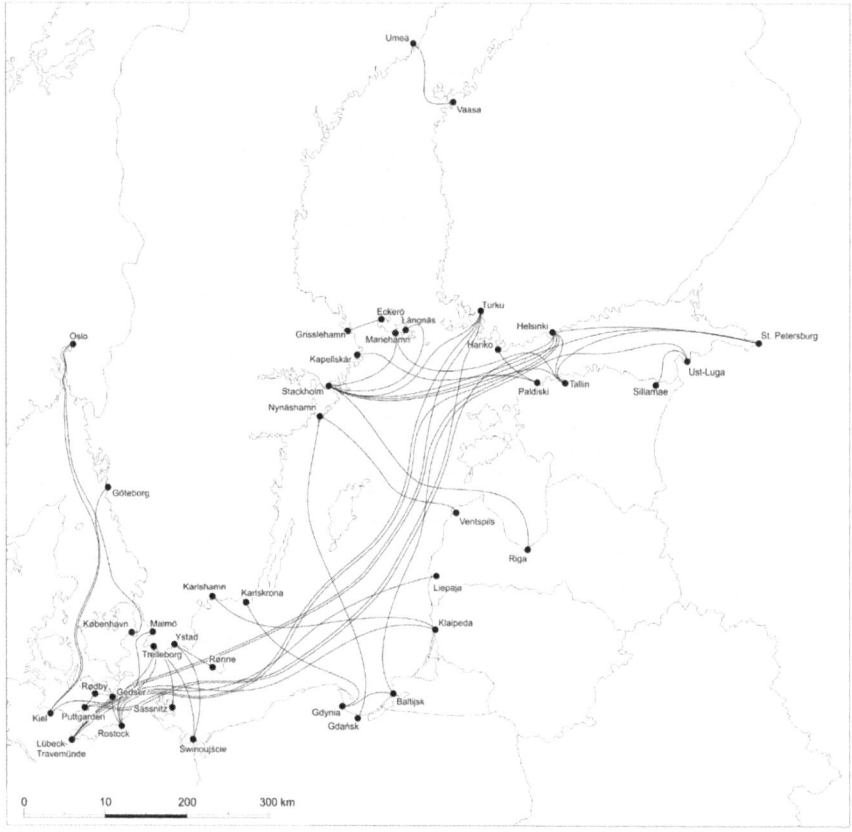

Fig. 1 Diagram of regular ferry routes on the Baltic Sea (except seasonal services, as on 31 August 2017). *Source* author's own materials

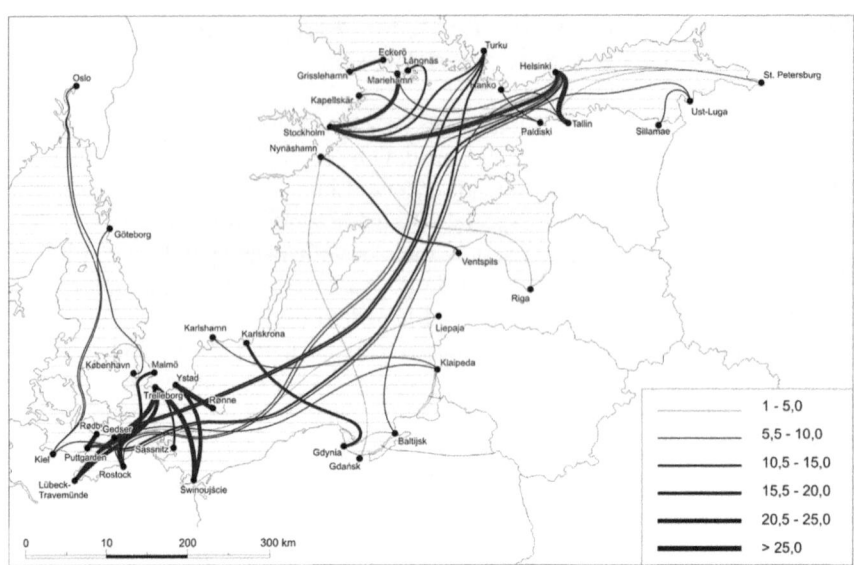

Fig. 2 Weekly service frequency on the main regular ferry routes on the Baltic Sea (except seasonal services, as on 31 August 2017). *Source* author's own materials

Sweden meet the 24-h return trip criterion, the most attractive proposition from the customer's point of view. Passenger traffic is largely affected by Poland and Sweden being EU members, hence the absence of duty-free shopping, so characteristic for Baltic ferry business (the Aland Islands in routes from Helsinki, Stockholm and Turku). The most tourist friendly policy is pursued by Stena Line, which for many years has been providing day trips and short cruises for tourists from Poland and Sweden. In recent years it has significantly broadened its offer for tourists from Sweden, supporting it with a successful advertising campaign resulting in growing sales of trips to Poland (Urbanyi-Popiołek 2016).

On the market of passenger service on regular ferry routes on the Baltic Sea, Polish routes play a moderately important role. The largest number or services is on the routes between Tallinn, Estonia and Helsinki, Finland—80 services a week. The traffic is largely generated by the price differences between the two countries and the very small distance—only 2 h at sea. Many services are also provided between German and Swedish ports for cargo traffic and between Sweden and Finland for tourist traffic (see Fig. 2).

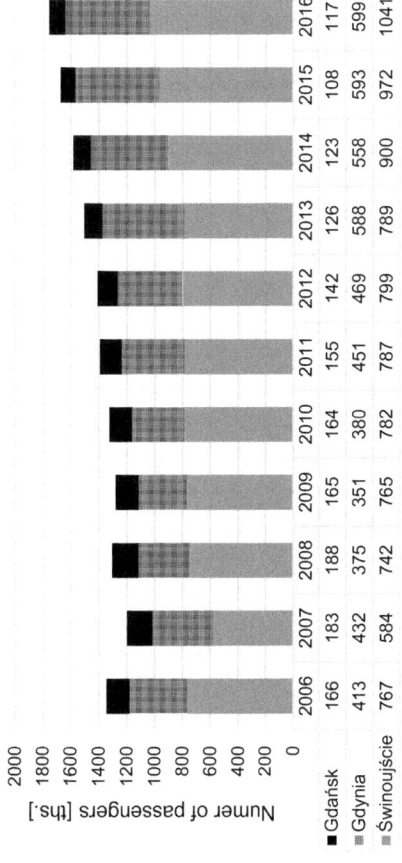

Fig. 3 Passenger service in Polish ports in regular ferry crossings without tourist cruises in 2006–2016. *Source* author's own calculation based on data provided by seaports in Gdańsk and Gdynia, Ferry Terminal Świnoujście and shipowners Polish Baltic Shipping, Stena Line, TT-Line and Unity Line; Christowa (2012)

The Volume of Services Provided by Polish Operators Compared to Other Baltic Sea Countries

The number of passengers carried by ferries on the Baltic Sea was rather stable in recent years at approx. 110–120 million passengers per year. The share of Poland in this market is relatively small at approx. 1% (Eurostat...). The number of passengers carried by shipowners operating out of Poland grew from 1200 thousand in 2007 to nearly 1800 thousand in 2016 (see Fig. 3). The largest volumes on the Baltic Sea are recorded in Denmark, Sweden and Finland. Services out of Estonian ports also play an important role. Shipping lines operating in Central Baltic have a smaller share in the market. A marginal role is played by services to Lithuanian and Latvian ports.

Among Polish ferry routes the most important one is from Gdynia to Karlskrona, where the number of passengers grew in recent years from 351 thousand in 2009 to 599 thousand in 2016. This route was particularly affected by the economic crisis in Europe in 2007–2009. A significant growth in passengers was also noted in the port in Świnoujście. The only route where the number of passengers went down is from Gdańsk to Stockholm conurbation, but it must be emphasised that Polish Baltic Shipping reduced the demand for that route in recent years by reducing service frequency (one operating ferry instead of two).

Summary

The coming years should bring about further increase in the volume of passengers and cargo moved between Polish and Nordic ports. Significant impulse will come from the completion of the southbound road infrastructure at national level (expressways S3, S7 and the A1 motorway). The completion of road building projects should encourage road hauliers from Central Europe (Czech Republic, Slovakia and Hungary) to use Polish ports. The future market will be strongly affected by economic, social and geopolitical factors (Matczak 2016). However, the key role should be played by economic/environmental aspects related to the need to replace the antiquated fleet. The EU Directive which obligates shipowners operating routes on the Baltic Sea to use desulphurized fuel (SECA zone) drives up the operating costs. Large financial outlays needed to build innovative ferries powered by LNG or other alternative fuels must be paid for by profits, which in case of passenger shipping are sensitive to the competition from low cost airlines. Good prospects for GDP in Poland and Sweden announced by the International Monetary Fund, and negative scenarios for such countries as Denmark, Estonia, Lithuania, Latvia or Germany should further intensify trading between Poland and Nordic countries, especially Sweden.

The ferry shipping market on the Baltic Sea may also be affected by the geopolitical situation, especially migration, which may lead to Schengen Zone being suspended. In this scenario trade could be significantly affected (Matczak 2016). Polish passenger ferry services are relatively stable. In recent years after 2008 constant growth is recorded on most routes. During this time shipowners also continued to increase supply by operating newer and larger ferries. Compared to the scale of the Baltic Sea market, services to and from Poland do not play the crucial role, but locally they are a significant stimulant for the economy, especially in tourism-connected sectors (Urbanyi-Popiołek 2012b). For the Polish shipowners (Polish Baltic Shipping and Polsteam), the key factor in short term perspective is the need to replace the antiquated fleet. This is related to the implementation of the government project *Batory Programme* (Program Batory...), which includes plans to build 10 new RoPax ferries. The success of this endeavour largely depends on establishing the financing and the ability to secure money from EU funds.

References

Christowa C (2012) Analiza i determinanty rozwoju przewozów promowych w basenie Morza Bałtyckiego ze szczególnym uwzględnieniem autostrady morskiej Świnoujście-Ystad. Prace Naukowe Politechniki Warszawskiej. Transport 82:7–24

Czermański E (2010) Żegluga promowa w regionie Morza Bałtyckiego w układzie Północ-Południe. Współczesna Gospodarka 1:69–81

Eurostat. http://ec.europa.eu/eurostat/data/database. Accessed 8 Sept 2017

Kotowska I (2010) Nowe koncepcje autostrad morskich w ramach środkowoeuropejskiego korytarza transportowego. Ekonomiczne Problemy Usług 49:165–178

Matczak M (2016) Polskie porty morskie w 2015 roku. Podsumowanie i perspektywy na przyszłość, Gdynia

Pacuk M (1997) Perspektywy polskiej żeglugi promowej w transbałtyckim systemie komunikacyjnym. Prace Komisji Geografii Komunikacji PTG 3:171–185

Polska Żegluga Bałtycka. http://www.polferries.pl. Accessed 8 Sept 2017

Port Gdańsk. http://www.portgdansk.pl. Accessed 8 Sept 2017

Port Gdynia. http://www.port.gdynia.pl. Accessed 8 Sept 2017

Program Batory – platforma rozwoju innowacji w polskiej gospodarce morskiej. http://www.gospodarkamorska.pl/Stocznie,Offshore/program-batory-%E2%80%93-platforma-rozwoju-innowacji-w-polskiej-gospodarce-morskiej.html. Accessed 30 Aug 2017

Stena Line. http://www.stenaline.pl. Accessed 8 Sept 2017

Termin Promowy Świnoujście. http://www.sft.pl/. Accessed 8 Sept 2017

TT-Line. http://www.ttline.com. Accessed 8 Sept 2017

Unity Line. http://www.unityline.pl. Accessed 8 Sept 2017

Urbanyi-Popiołek I (2010) Polskie porty w koncepcji autostrad morskich. Ekonomiczne Problemy Usług 49:217–224

Urbanyi-Popiołek I (2012a) Wpływ kryzysu ekonomicznego na rynek żeglugi promowej na Morzu Bałtyckim. Logistyka 2:287–292

Urbanyi-Popiołek I (2012b) Funkcje gospodarcze żeglugi promowej na przykładzie linii promowych z portów Trójmiasta. Logistyka 5:772–778

Urbanyi-Popiołek I (2016) Na bałtyckim rynku promowym. https://www.namiary.pl/2016/03/09/na-baltyckim-rynku-promowym/. Accessed 10 Aug 2017

The Kaliningrad Region in the System of Global Transport Corridors: On the Prospects of the Region's Development as a Transportation Hub

Ivan S. Gumeniuk⬤ and Lidia G. Gumeniuk

Abstract

Being the mainstay of intraregional freight and passenger traffic and international trade conducted by Russian economic entities, the Kaliningrad transport network is seeking its place in the transit industry. The regional transport network's potential to become a transportation hub will largely depend on the implementation of China's Belt and Road Initiative, which is designed to create a single trade corridor carrying east-west commodity traffic. Today, there are four global projects aiming to create a transport link between Asia-Pacific (first of all, China) and the countries of Western Europe. In this article, we analyse the feasibility of proposed transportation projects and consider their effect on the Kaliningrad region's transport network. Our methodology rests on an analytical comparison of the projects from the perspective of their practicality and completion periods. Our findings suggest that the implementation of certain projects will impede the region's integration into global transport corridors. However, it is very unlikely that these projects will be completed in a mid-term perspective. It is more realistic to expect the implementation of transportation projects that treat the Kaliningrad region as an important transportation hub.

Keywords

Global transport corridors · Transport projects · Transit potential · Kaliningrad region

I. S. Gumeniuk · L. G. Gumeniuk (✉)
Institute of Environmental Management, Urban Development and Spatial Planning, Immanuel Kant Baltic Federal University, Kaliningrad, Russia
e-mail: LOsmolovskaya@kantiana.ru

Introduction

The transport network of the Kaliningrad region caters for a broad range of intraregional and national freight and passenger traffic needs. Thanks to its geographical position, it has a potential to become a transportation hub supporting freight traffic between the world's two largest economic poles—Asia-Pacific with its centre in the People's Republic of China (PRC) and Western Europe (the European Union). A global trend that will affect not only the development of international transport corridors on the Eurasian continent but also the architecture of relations between countries is the PRC's Belt and Road Initiative (Avdokushin 2015). This global economic initiative introduced by the PRC in 2013 is aimed to create a global trade corridor carrying direct east-west commodity traffic. This economic corridor will connect Asia-Pacific in the east and the developed countries of Europe in the west. Home to over three billion people, the countries engaged in the project have a total GDP of approximately USD 21 trillion. The project was aimed at simplifying trade rules (to eliminate trade barriers) between the participant countries, increasing the quality and swiftness of economic transactions in the region, improving the transboundary transport infrastructure, and constructing new high-capacity transport corridors on the territories of the participant states (The New Silk Road and Its Importance for Russia 2016).

At first, the project's infrastructural component consisted of the technological and infrastructural convergence of two transport corridors—the Silk Road Economic Belt and the Maritime Silk Road. Project participants are competing for USD 40 billion worth of Chinese investment (The New Silk Road, or How China Desires to Unite All 2017). This Chinese initiative enlivened geopolitical competition for the right to cater for the gigantic freight traffic from China and other countries of Asia-Pacific to the EU (Kosorukov and Barto 2016).

Today, most of the EU—Chinese trade is carried by sea along the Maritime Silk Road. UNESCO estimates the Asia–Europe–Asia container traffic at 22 million TEUs (14.9 million TEUs from Asia to Europe and approximately 6.8 million TEUs from Europe to Asia) (Review of Maritime Transport 2016). However, growing geopolitical risks ('Somalian pirates', local armed conflicts in the Persian Gulf and the Suez Canal area, etc.) and geoeconomic costs (the need to increase imports against the background of higher environmental levies and freight costs (Bezborodov 2017), etc.) make China seek new ways for land corridors to deliver the required traffic capacity. We will study three the most mature projects, which can meet China's growing needs for transport corridors. We will look into each project in detail and focus on its effect on the prospects of the Kaliningrad region to become a transportation hub.

The Eurasian Railway Corridor Bypassing Russia

The idea of establishing a rail link between China and the EU bypassing Russia grew out of the TRACECA—Transport Corridor Europe-Caucasus-Asia—project (Official website of the TRACECA project 2017). Launched by the EU and the partner countries, the project aimed to create a Europe–Caucasus–Asia transport corridor within the international cooperation programme. Later, new participants joined the project, which grew to involve the railway section of Pan-European Corridor IX linking Ukraine's seaports to those of Poland and Lithuania. In terms of organisation, the European part of the corridor performs properly. The Viking train runs between Odessa and Klaipeda (Official website of the VIKING Train project 2017). However, the Ukraine—China leg of the route is still being tested. In January 2016, a Ukraine–Georgia–Azerbaijan–Kazakhstan–China train embarked on a test voyage to reach its final destination in the village of Dostyak at the Chinese-Kazakhstani border. Due to a range of different circumstances, the voyage took 15.5 instead of 11–12 days.

At the stage of feasibility studies and after the test voyage, many experts questioned the economic practicality of the railway route, two sections of which— across the Black and Caspian Seas—are spanned by ferry links. Such a structure of the route translates into four port calls. Special equipment is required to board the ferry. Moreover, high risks are associated with adverse weather conditions at sea, particularly, in winter. All this means higher freight transportation costs and longer rail journey times.

The effect of the project on the prospects of the Kaliningrad region's transport network. Despite the project's low efficiency in terms of economics and logistics, apparent political will may expedite the delivery of the project. If the project is successfully completed, major freight commodity traffic between China and the EU will bypass Russia in general and the Kaliningrad region in particular. In the best case, the Kaliningrad region will perform support functions and 'intercept' part of the freight traffic shared between Poland's and the Baltics' seaports. Under the most optimistic scenario, the Kaliningrad railways are not likely to handle more than 10–15 thousand TEUs. Since the maximum gross mass of one TEU is 24,000 tonnes, the annual tonnage will not exceed 0.5–1 million tonnes. Thus, the completion of this project will deny the Kaliningrad region almost any chance of becoming a transportation hub.

The Project of an Integrated EAEU Transport Corridor—Kazakhstan, Russia, and Belarus

In 2014, the joint actions of EAEU countries resulted in the establishment of the United Transport and Logistics Company (Official site of OTLK 2017). The company was co-founded by the Russian Railways, the Belarusian Railway, and Kazakhstan Temir Zholy. OTLK acts as an operator that provides infrastructure for the partner countries' transit railway traffic. The train travels 5.5 days to cover the 5430 km between the terminal points of the route—Belarus's Brest and Kazakhstan's Dostyk. Despite the short project timeline, OTLK secured a strong position in the transit traffic market. Experts estimate the total 2017 transit tonnage carried along the China–Europe–China route at 300–350 thousand TEUs. In a short-term perspective, it can reach 500 thousand TEUs per year.

The transit corridor is associated with not only a growing transit traffic but also an expanding geography of transportation hubs involved. For instance, the corridor infrastructure made a railway link between China and the UK possible. A freight train left the Chinese city of Yiwu in the Zhejiang province on January 2, 2017, and arrived at Barking Rail Freight Terminal in London seventeen days later (The train from China for the first time in history arrived at London station 2017). London became the fifteenth European destination of Chinese freight trains. The proposed railway route to London creates a link that is thrice as fast as maritime shipping at one-fifth of the cost of transportation by air.

The effect of the project on the prospects of the Kaliningrad region's transport network. The development of freight traffic along this transport corridor is already affecting the Kaliningrad railway network. In September 2017, the first container train travelling from Europe to China arrived at the local town of Chernyakhovsk from Poland's city of Łódź. Forty-one FEUs were transhipped from platforms suited for the 1435 mm 'European' gauge to those designed for the 1520 mm 'Russian' gauge. The train was bound to China's Chengdu (The OTLC organized a regular transit of containers between Europe and China through the Kaliningrad region 2017). At the first stage of the project, the Kaliningrad railway terminals are expected to handle up to 365 container trains. This number may grow later. If the project's infrastructure proves itself capable of handling the freight transit between the EU and China, the Kaliningrad railways will enjoy sufficient demand for freight transit, partly thanks to 'hosting' the interconnection of two gauge standards.

The Northern Sea Route Project

An emerging alternative to both the principal maritime transport corridor via the Suez Canal and the above railway corridor is the Northern Sea Route (NSR). Despite the rapid infrastructure enhancement of the Northern Sea Route and projects aiming to modernise the fleet, the NSR as a major route carrying the transit traffic between the EU and Asia-Pacific—particularly, China, South Korea, and Japan—remain a long-term prospect, which may become reality only after 2030.

In a short-term perspective, the NSR aims to meet national needs. Any transit traffic along the route will be a 'pioneering' project. However, after 2025, the configuration of global transport corridors—in a combination with mature infrastructure and functioning logistics chains—may result in the NSR handling up to 40% of the transit traffic between the EU and Asia-Pacific.

The effect of the project on the prospects of the Kaliningrad region's transport network. The main corridors carrying transit freight along the Northern Sea Route bypass the Kaliningrad region. Thus, from the perspective of the region's development as a transportation hub, this project is the least beneficial.

Overall, there are four global projects aimed to support traffic between Asia-Pacific and Western Europe. One might consider the use of the Trans-Siberian Railway as transit infrastructure one of such projects. However, the specifications of the TSR make it suitable for carrying only national traffic between Russia's European and Far Eastern regions. Without a comprehensive infrastructure modernisation, this transport corridor will remain an exclusively national thoroughfare.

Table 1 demonstrates the systematization of these projects based on their impacts on the transport network of the Kaliningrad region.

Table 1 Projects aimed to create transport corridors between Asia-Pacific and Western Europe and their effect on the prospects of the Kaliningrad region's transport network

No.	Project	Brief description	Full capacity deadline	The project's effect on the transport network of the Kaliningrad region
1	Maritime Silk Road (via the Suez Canal)	The project is underway. As of 2015, 22 million TEUs were shipped along the Europe–Asia route	The corridor has reached its full capacity.	'0' (zero effect on the Kaliningrad region transport network)
2	Eurasian Rail corridor bypassing Russia	A railway route via Ukraine, Georgia, and Kazakhstan	Test voyages have been made. Deadlines are not clear	'+' (minimum positive effect on the Kaliningrad region of the project is completed)
3	Integrate EAEU transport corridor	The project has been carried out since 2013 by OTLK—a company established by Russia's Belarus's, and Kazakhstan's national railways	The project has been rapidly developing since 2013. It is expected to reach full capacity by 2025	'++' maximum positive effect on the Kaliningrad region. If the project is completed, the region will be included in the logistics chains of the transport corridor
4	Northern Sea Route	The route's infrastructure is undergoing the process of modernisation	After 2030	'—' negative effect on the Kaliningrad region. If the project is completed, the region will be 'isolated' from logistics and transport corridors

(continued)

Table 1 (continued)

No.	Project	Brief description	Full capacity deadline	The project's effect on the transport network of the Kaliningrad region
5	Trans-Siberian Railway as a transit corridor	The project is under expert evaluation	No data	'0' (zero effect on the transport network of the Kaliningrad region)

Conclusion

The project of an integrated rail transport corridor, which was initiated by the EAEU member states is associated with the greatest positive effect on the prospects of the Kaliningrad region to become part of international transport corridors. The other projects proposed are less beneficial for the region. However, some of them may contribute to the transit function of Kaliningrad. Russia's chances to unlock its potential of a transit country will depend on the fact whether the above projects enjoy geopolitical support. The geopolitical factor will also affect the development of regional transport networks.

References

Avdokushin EF (2015) Tigr prygnul, drakon vzletel – proekt «Odin poyas - odin put'» (The tiger jumped, the dragon took off—the project "one belt—one way"). Theory Pract Issues New Econ 4(36):4–17 (in Russ.)

Bezborodov A (2017) For the first time in 30 years ocean freight is more expensive than land transportation in Eurasia. http://eurasia.expert/chto-zhdet-sovmestnyy-zheleznodorozhnyy-proekt/. Accessed 12 June 2017 (in Russ.)

Kosorukov AA, Barto EV (2016) Velikii shelkovyi put' v konkurentnom prostranstve mirovoi politiki (The great silk road in the competitive space of world politics). Natl Secur 2:191–211 (in Russ.)

Manukyan Zh (ed) (2017) Novyi shelkovyi put', ili Kak Kitai khochet vsekh ob"edinit' (The New Silk Road, or how China wants to all). https://ria.ru/economy/20170513/1494227526.html. Accessed 10 June 2017 (in Russ.)

Official site of OTLK. http://www.utlc.com/routes. Accessed 10 June 2017

Official website of the TRACECA project. http://www.traceca-org.org/en/home/. Accessed 10 June 2017

Official website of the VIKING Train project. http://www.vikingtrain.com/about. Accessed 10 June 2017

Petrovsky V, Larina A, Safronova E (2016) Novyi Shelkovyi put' i ego znachenie dlya Rossii (The New Silk Road and its importance for Russia). DeLi Plus, Moscow (in Russ.)

Review of Maritime Transport (2016) United nations conference on trade and development. http://unctad.org/en/PublicationsLibrary/rmt2016_en.pdf

The OTLC organized a regular transit of containers between Europe and China through the Kaliningrad region. http://www.utlc.com/news/tranzit-kaliningrad. Accessed 12 June 2017 (in Russ.)

The train from China for the first time in history arrived at the London station. https://meduza.io/news/2017/01/18/poezd-iz-kitaya-vpervye-v-istorii-pribyl-na-vokzal-londona. Accessed 10 June 2017

Forms and the Role of Roadside Memorials in the Gdańsk Agglomeration, Poland

Lucyna Przybylska ⓘ

Abstract

The paper fills in a gap in the literature on roadside memorials in Poland. The aims of this article are to find a typical form of a roadside memorial in the Gdańsk agglomeration (northern Poland) and to explore the role (function and meaning) of roadside memorials. The results were achieved through the field studies and a survey conducted in 2017. The first part of the article provides a description of investigated remembrance sites, while the other one analyzes the opinions of students of the University of Gdańsk. A typical roadside memorial is a metal, rusty Latin cross, parallel to the road, decorated with artificial roses and accompanied with one burnt out votive candle. Nearly all the students are of the opinion that Poles put up memorial crosses in order to commemorate their family members who died in tragic road accidents. About one third of the respondents believe that memorial crosses serve the purpose of warning other drivers about a dangerous fragment of the road. The survey points to a slightly larger percentage of the religious than the cultural argumentation as regards the choice of the cross instead of other objects to mark accident sites in Poland.

Keywords

Gdańsk · Memorial crosses · Poland · Roadside memorials

L. Przybylska (✉)
Department of Spatial Management, Faculty of Oceanography
and Geography, University of Gdańsk, Gdańsk, Poland
e-mail: lucyna.przybylska@ug.edu.pl

© Springer Nature Switzerland AG 2020   217
G. Fedorov et al. (eds.), *Baltic Region—The Region of Cooperation*,
Springer Proceedings in Earth and Environmental Sciences,
https://doi.org/10.1007/978-3-030-14519-4_24

Introduction

The paper fills in a cognitive gap in the literature on roadside memorialisation in Poland. A roadside memorial is a set of various objects, usually including crosses, flowers and votive candles, placed at a particular site to commemorate a victim of a fatal road accident. A number of research on roadside memorials has increased recently (a few authors publishing in the 1990s and over 20 in the XXI century). As for the spatial scope of the research by continents, the majority of authors conducted their studies on roadside memorials in North America (Bednar 2013; Clark and Cheshire 2004; Dickinson and Hoffmann 2010; Everett 2002; Henzel 1991; Owen 2011; Reid and Reid 2001; Tay et al. 2011; Zimmerman 1995), followed by Europe (Diasio 2011; Klaassens et al. 2009; Maddrell 2013; Nešporová and Stahl 2014; Petersson 2009; Przybylska 2015), Australia (Breen 2006; Brien 2014; Clark and Cheshire 2004; Clark and Franzmann 2006; Hartig and Dunn 1998; Welsh 2017) and Asia (Cohen 2012). In Europe, the studies on the topic were conducted in Sweden, Great Britain, the Netherlands, Italy, Romania, the Czech Republic and Poland. Most works are focused greatly on the forms and functions of roadside memorials (Diasio 2011; Everett 2002; Petersson 2009); studies of different types concern the law, official policy and social perception (Bednar 2013; Dickinson and Hoffmann 2010; Reid 2014) as well as the impact of roadside memorials on drivers' behaviour and safety (Tay et al. 2011). The paper is a part of the predominant way of examining roadside memorials by looking for their detailed characteristics (forms) and the role (function and meaning) they play. It must be emphasised that some features of commemoration sites presented in the paper are often overlooked in the relevant papers (e.g. colours of votive candles and flowers) or they are only mentioned to be present but lack in quantitative results.

The aims of this article are to find a typical form of a roadside memorial in the Gdańsk agglomeration (northern Poland) and to explore their role. They were achieved through the field studies (in July 2017) and a survey (in April and June 2017). The study was carried out on a 50 km stretch of the national road No. 20, which runs close to western administrative border of the city of Gdańsk and further south-west to the Lakelands. A one-page anonymous questionnaire had been distributed among students of the University of Gdańsk. The choice of the route and the questions in the survey resulted from the assumptions of the research project, which has been implemented since January 2017. The aim of the project is to describe the form and function, and to explain the meaning of memorial crosses along public roads in Poland. The main research questions of the project are the following: Why are crosses placed at accident sites in Poland? By whom, and to what extent is a memorial cross associated with religion and to what extent with other aspects of meaning? The description of the features of 40 commemoration sites, presented further in the first part of the article, situated in a selected part of the road, as well as the opinions of 88 students, are the first outcomes of a 3-year-long project. It must emphasised that the terms "roadside memorials" and "commemoration sites" are used interchangeably in the paper.

The sample of the survey is neither representative for general opinion nor for students of the Gdansk University. However, it is useful to fulfill this paper`s scope to explore the role of roadside memorials and test questions for all Poland public survey scheduled in future. The author is aware that in order to understand the meaning of commemoration sites it is not enough to examine only students. It is also necessary to identify the meaning attached to these memorials by those who created them. This will be the next step of the ongoing project.

The Location of Roadside Memorials

The research identified two types of 40 roadside memorials, considering their location with respect to the road and the traffic participants in motion: either parallel or perpendicular. The great majority of memorials are parallel to the road (29, which means nearly ¾)—in order to have a full view of a memorial cross or a memorial plate, we must stand parallel and at the same time with our back turned to the road. On the other hand, 10 roadside memorials (i.e. ¼) are situated as if they were "looking" at the approaching vehicle. One memorial can be included in both types distinguished above, due to the concentric shape of the lower elements building up this commemoration site, i.e. a number of votive candles and some flowers laid on the grass. Roadside memorials are situated, on average, every 1.2 km along the studied road. Most (27) commemoration sites were found by a straight stretch of the road, about ¼ near a turn (9), only 3 at the exit from a road without right of way, and one such memorial was located at a roundabout.

Objects Found at Commemoration Sites

As many as 38 out of 40 commemoration sites contain a cross (95%). This very high rate corresponds to memorial crosses rate in Romania (98%) (Nešporová and Stahl 2014) and differs from results in other countries (Cohen 2012; Dickinson and Hoffmann 2010; Klaassens et al. 2009; Welsh 2017). Two roadside memorials without a cross are vertical gravestones with votive candles and flowers. In total 38 roadside memorials contain 42 crosses. The majority of these memorials feature one cross (35). Only two commemoration sites contained two and one site—three memorial crosses.

About half of sites include votive candles or flowers (Table 1). Moreover, it is quite common for a site to contain both, flowers and votive candles (16), or no votive candles or flowers at all (9). All in all, 61 votive candles were counted, placed from 1 to 6 at 25 roadside memorials, with only one votive candle burning on the day of inventory. Both commemoration sites without memorial crosses had a larger number (5) of votive candles. One can usually find one votive candle—at 9 sites with memorial crosses or two candles—at 7 such places. The votive candles are more and more often found standing in the upright position (45, i.e. ¾) than lying on the side (16). Glass votive candles are more common (52) than plastic (7) or metal-shielded (1) ones; the material which one candle was made of was not recorded. The votive candles are of different sizes and ornamentation. The colours of the candles were noted down: over half of them were transparent (36), every sixth one was red (10). Other colours included white (8), pearl (2), pink (2), golden and blue (1). Transparent, white and red candles are easily available in the shops. The votive candles usually stand on the ground, occasionally on a concrete or stone panel situated on the ground (3 roadside memorials). It also happens that a votive candle is placed on the cross, on a special metal stand (4 sites).

There are flowers (mostly artificial) at half of the roadside memorials (21). Real plants were present only at two sites (pansies and marigolds in flower pots, thuja twig). Flowers can be placed on the memorial cross (11 sites) or/and on the ground (10 sites)—in both cases, in or without a pot. The flowers are arranged in bouquets of different sizes; a wreath was found only once. Artificial flowers imitate 14 different flower species of different colours (Table 2). Similarly to field study in Texas roses appeared to be the most popular flower (Everett 2002). Carnation was the second-most utilized flower in Texas and chrysanthemum in the Gdańsk agglomeration. The difference is also visible in colours. Red roses and carnations prevail in the US and pink roses and yellow chrysanthemum in Poland.

On five crosses, other objects typical of remembrance sites included a small metal plate with letters INRI inscribed (a Latin acronym customarily placed on crucifixes, standing for Iesus Nazarenus Rex Iudaeorum). Four roadside memorials had written information about the deceased or about the accident which is a very low rate contrary to more than 40% roadside memorials with inscriptions to be reported in the US (Everett 2002; Owen 2011) and Australia (Clark and Cheshire

Table 1 Votive candles and flowers at commemoration sites

Presence	Votive candles	Flowers
Present	25	21
Absent	15	19
Exclusively	9	5

Source Author's elaboration

Table 2 The most common flowers at commemoration sites

Species	Colour
Rose (10)	Pink (4), red (3), purple (2), tea (1)
Chrysanthemum (7)	Yellow (3), purple (2), white (1), pink (1)
Daffodil (5)	Yellow
Tulip (4)	White (2), yellow (2)
Iris (3)	Purple
Calla lily (2)	Pink

Source Author's elaboration

2004; Welsh 2017). At three of them, a concrete or stone slab was found, probably placed there as a stand for the votive candles. At two places, a gravestone (vertical and horizontal) was found and at other two—a plate on the cross (one new with inscriptions and the other one blank—the inscriptions had probably faded). Each of the following objects were found once at the commemoration sites: a motorbike helmet, bead necklace, a boulder, a small teddy-bear, a Virgin Mary figurine, a metal oak twig, and a small crucifix on a large memorial cross.

Description of Memorial Crosses

At 38 commemoration sites with memorial crosses, most crosses (34) were inserted directly into the ground and presented as single standing crosses or groups of standing crosses. Only at three sites, they were "lying crosses", and at one—the cross was fixed to a wall. Only two types of material were observed, which the crosses were made of: metal or wood. There are definitely more metal crosses (36) than wooden ones (6). Metal is a long-lasting, cheaply available and durable material which also prevail in Romania (Nešporová and Stahl 2014) but is rare in the US (Everett 2002; Zimmerman 1995). Nearly half of the roadside memorials with crosses (18 out of 38) had a figure of Jesus placed on the intersection of the two beams of the cross. Among 42 crosses standing at 38 sites, the smallest one was 67 cm tall, while the biggest—152 cm (the arithmetic mean—102 cm). Both extreme cases are wooden crosses.

We can mostly find rusty crosses (20): brown or brown-and-black, colour brown coming from the rust. The rusting of the metal memorial crosses seems to prove a long history of the cross standing at a given site, i.e. a long time of exposure to the weather conditions, which cause natural corrosion of metal. At the same time, the presence of rust may be due to the lack of or insufficient metal conservation during exposure. The colour defined as "shiny metallic" most probably regards relatively new crosses, not only still untouched by rust but even shining on sunny days. They are easily discernible for the participants of road traffic (Table 3).

Table 3 The colours of memorial crosses

Wooden cross		Metal cross	
Colour	Number	Colour	Number
Black	2	Black	8
White	1	White	1
Rotten wood	2	Rusty	20
Natural wood	1	Metallic, "shiny"	5
		Grey, "matt"	2
Total	6	Total	36

Source Author's elaboration

All the memorial crosses except one are Latin crosses, with one horizontal beam. The exception is the Russian Orthodox cross (with two horizontal beams). Considering the shape of the crosses, we should notice that half of them are ornamented and the other ones are plain. The majority of the plain crosses are made of metal; they stand at 14 commemoration sites (1/3). It was noticed that the figure of Jesus is most often found on wooden crosses; 4 out of 6 such crosses have the figure of Jesus. This fact can be referred to the Biblical account of Jesus' death on a wooden cross and the fact that it is easier to attach the figure to wood than to metal.

It was noticed that the shapes of the ornamented crosses vary in details. The exceptions are three crosses of nearly identical shape: the beams are surrounded with parallel bars; in each quarter of the cross there are three rays radiating from the central figure of Jesus, the ends of the beams are bevelled at the angle of 45°, and there is a rounded rooflet over the figure of Jesus (in one case there is no rooflet, but we can see three holes which probably remained after fixing a rooflet earlier). These crosses are heavily rusted; they are not situated within the area of one village, but it seems obvious that they were designed by the same person.

Students' Perception of Memorial Crosses

The perception of memorial crosses was evaluated on the basis of a survey conducted among 88 MA students (aged around 22), attending classes at the Institute of Geography, Gdańsk University. The group of respondents included 37 students of spatial management and 51 students of geography and biology; 59 women and 29 men. In response to the first question of the questionnaire: Have you heard of crosses standing at accident sites by a road?, all respondents but one person, who ticked both "yes" and "no" (it was considered to be an invalid answer), answered affirmatively. Thus, young people are commonly familiar with the phenomenon of memorial crosses. The second question was: Should memorial crosses along the road be removed? The students could choose the same options (Table 4) which journalists in agreement with the Gdańsk City Hall had offered to Internet users six years earlier (Przybylska 2015). It created a possibility of an interesting comparison. Firstly, the 88 students are less in favour of removing memorial crosses (33.5%) than the group of nearly 4000 Internet users taking part in the survey in 2011 (48%). Secondly, both the young people and the Internet users agree that memorial crosses should not be removed, because they are powerful warning signs for drivers (44% of indications in each group), and that a road is not a proper place for a cross (19% indications in each group).

The questionnaire closes with a question about faith in God (I strongly believe, I believe, I don't believe) and whether the cross is very important, important or not important in their lives. It turned out that most people are believers (63%) and those for whom the cross is important (56%). Strong faith and a very important role of the cross in life were declared by 15% of the students. The non-believers in this group made up 22% and for 29%, the cross is not important. A relationship was sought

Table 4 Answers to the survey question: Should memorial crosses along the road be removed?

Answer options	Students (%)	Internet users (%)
Yes, because they distract the drivers and have been put up illegally	14.5	29
Rather yes, a road is not a proper place for a cross	19	19
Rather not, because they have become commonplace anyway	23	8
No, they are a powerful warning sign for drivers	43.5	44

Source Author's elaboration based on students' questionnaires and Przybylska (2015)

between those indicators and the respondents' sex with regard to 28 people, who declared that they wanted memorial crosses to be removed from the roads (answers "yes" and "rather yes" in Table 4). It turned out that female/male variable did not determine those answers. On the other hand, the majority of those who wanted the crosses to be removed from roads were people who did not believe in God and those to whom the cross was not important in their lives.

Two key open questions regarding the perception of the motivation to put up memorial crosses were placed in the middle section of the questionnaire. Open, longer or shorter answers provided by the students were grouped by similarity into several topical categories. Nearly all students (71 out of 88) relate memorial crosses to the commemorating function. They think that Poles put up the crosses in order to commemorate a victim of an accident, family members who died in a tragic accident or to commemorate the site of death or accident of their close relatives. Almost 1/3 of the students (26) quoted the argument of warning other drivers about a dangerous section of the road. It is interesting that in this group, only 6 respondents wrote only about the warning function of memorial crosses, and the remaining 20 persons identified the commemorating function as well. Only 5 people explained that the custom of putting up memorial crosses is rooted in religion, religious tradition or history.

Nearly half of the students (40) quoted only religious arguments, 1/3—only cultural ones, and 12 people—both, when answering the question why Poles put up memorial crosses at road accident sites, and not something else, such as a boulder or just flowers. Among 60 various answers counted as religious arguments, 38 pointed to faith or religion, or the cross as a symbol of faith, religion, Christianity or Catholicism, 17 people wrote about Poland as a Catholic or Christian country, and only 5 people explained the phenomenon with Poles' "religiousness". Among the cultural arguments (49), there were 21 statements pointing to the cross as a symbol of death or clearly associated with death, and 11 answers referring to the custom of burying the dead at cemeteries, with crosses standing next to graves. Moreover, 8 respondents explained the phenomenon of memorial crosses with tradition and culture, not using adjectives tinted with religious meaning, while the remaining ones wrote about commemorating dead people, a "commemorative symbol" or an "important symbol"—those opinions were also regarded as cultural arguments.

Conclusions

To sum up, considering the occurrence of the measured features of 40 roadside memorials situated along road No. 20 in the Gdańsk agglomeration, it can be concluded that a typical roadside memorial is a rusty, metal, single Latin cross, 102 cm in height, placed parallel to the road and 3 m from its edge; at the foot of the cross, there is one burnt out votive candle, and the cross is decorated with artificial roses. The results of field research indicate that the forms of spontaneous roadside memorialisation reflect the denominational structure of Polish society. The high percentage of memorial crosses (95%) corresponds to the percentage of the adherents of the Roman Catholic Church in Poland (94.3%) and in Gdańsk Metropolis in 2015 (93.6%) (Statistical Yearbook of the Republic of Poland 2016). Other measured features of commemoration sites (especially flowers, candles and crosses) seem to be of the same kind as those put near headstones in cemeteries in Poland. A separate comparative study on this matter is needed. The parallelism between objects left at gravesites and objects left at roadside memorials is documented in research papers internationally (Welsh 2017), but not examined in Poland.

Nearly all the students included in the study are of the opinion that Poles put up memorial crosses in order to commemorate close relatives who died in a tragic road accident. Nearly 1/3 of the students said that the crosses are put up to warn other drivers about a dangerous section of the road. These results are similar to Klaassens et al. (2009) findings, according to which the basic function of roadside memorial crosses is commemorating victims of fatal accidents and, secondly, warning other road users. Furthermore, the survey conducted among students indicates a slight advantage of the religious argumentation over the cultural one when it comes to explaining the choice of the cross and not other objects to mark the site of a road accident in Poland.

The authors sees the survey presented above as a pilot study before representative research planned for whole Poland. The obtained results should be compared with the outcomes of individual interviews with people who have put up memorial crosses. Do they also show that the commemorative and warning roles of the memorial crosses predominate and that the religious motivation is only slightly stronger than the cultural one? The above description of the forms and students` perception of roadside memorials opens the presentation of the study results for the international audience. In the future, as the project advances, the conclusions will be confronted with analogous research in other regions of the country, which the author believes will be a significant contribution to the development of interdisciplinary research on roadside memorials.

Acknowledgements This paper is a part of the research project, "Memorial crosses along Polish roads". It has been sponsored by National Science Centre, Poland according to decision number DEC-2016/21/B/HS1/00823.

References

Bednar RM (2013) Killing memory. Roadside memorials removals and the necropolitics of affect. Cult Polit 9(3):337–356

Breen LJ (2006) Silenced voices. Experiences of grief following road traffic crashes in Western Australia. Dissertation, Edith Cowan University

Brien DL (2014) Forging continuing bonds from the dead to the living: Gothic commemorative practices along Australia's Leichhardt highway. J Media Cult 17(4). http://journal.mediaculture.org.au/index.php/mcjournal/article/view/858. Accessed 8 Sept 2017

Central Statistical Office (2016) Statistical yearbook of the Republic of Poland. http://stat.gov.pl/obszary-tematyczne/roczniki-statystyczne/roczniki-statystyczne/rocznik-statystyczny-rzeczypos politej-polskiej-2016,2,16.html. Accessed 18 Sept 2017

Clark J, Cheshire A (2004) RIP by the roadside: a comparative study of roadside memorials in New South Wales, Australia, and Texas, United States. Omega J Death Dying 48(3):203–222

Clark J, Franzmann M (2006) Authority from grief, presence and place in the making of roadside memorials. Death Stud 30(6):579–599

Cohen E (2012) Roadside memorials in Northeastern Thailand. Omega J Death Dying 66(4): 342–363

Diasio N (2011) Pamięć nietrwała. Estetyzacja żałoby prywatnej w przestrzeniach publicznych. In: Borzyszkowski J (ed) Nekropolie Pomorza, Nadbałtyckie Centrum Kultury, Gdańsk, pp 619–628

Dickinson GE, Hoffmann HC (2010) Roadside memorial politics in the United States. Mortality 15 (2):154–167

Everett H (2002) Roadside crosses in contemporary memorial culture. University of North Texas Press, Denton

Hartig KV, Dunn KM (1998) Roadside memorials: interpreting new deathscapes in Newcastle, New South Wales. Aust Geogr Stud 36(1):5–20

Henzel C (1991) Cruces in the roadside landscape of Northeastern New Mexico. J Cult Geogr 11 (2):93–106

Klaassens M, Groote P, Huigen PP (2009) Roadside memorials from a geographical perspective. Mortality 14(2):187–201

Maddrell A (2013) Living with the deceased: absence, presence and absence-presence. Cult Geographies 20(4):501–522

Nešporová O, Stahl I (2014) Roadside memorials in the Czech Republic and Romania: memory versus religion in two European post-communist countries. Mortality 19(1):22–40

Owen M (2011) Louisiana roadside memorials: negotiating an emerging tradition. In: Santino J (ed) Spontaneous shrines and the public memorialization of death. Palgrave Macmillan, London, pp 119–145

Petersson A (2009) Swedish offerkast and recent roadside memorials. Folklore 120:75–91

Przybylska L (2015) Memorial crosses in Poland: a commonplace and contested element of public roads. Geografie 120(4):507–526

Reid A (2014) Private memorials on public space: roadside crosses at the intersection of the Free Speech Clause and the Establishment Clause. Nebr Law Rev 92(1):124–184

Reid JK, Reid CL (2001) A cross marks the spot: a study of roadside death memorials in Texas and Oklahoma. Death Stud 25:341–356

Tay R, Churchill A, Barros AG (2011) Effects of roadside memorials on traffic flow. Accid Anal Prev 43(1):483–486

Welsh SM (2017) Private sorrow in the public domain: the growing phenomenon of roadside memorials. Dissertation, Charles Sturt University

Zimmerman TA (1995) Roadside memorials in five south central Kentucky counties. Dissertation. Western Kentucky University

The Modern Landscapes of the Baltic Macroregion and the Role of Socioeconomic Factors in Their Development

Elena A. Romanova ⓘ

Abstract

Human occupation, along with natural factors, is a major determinant of the modern landscape genesis on populated territories. In this article, I examine the modern landscapes of the Baltic macroregion. I use a unique method of landscapes zoning. This method consists in the calculation of a synthetic measure—the indicator of the effect of socioeconomic factors on the modern landscape genesis, i.e. of the degree of anthropogenic differentiation of landscapes. It is effective for any territorial unit (NUTS 2 and 3 for the EU countries and the level of a region for Russia). The indicator consists of five components: the average population density, the forest coverage and agricultural land use, the proportion of lands allocated to construction and roads, and the proportion of uncultivated land (bare rocks, sand, wetlands, glaciers, etc.). This study has practical relevance, since it offers a new method of landscape zoning that takes into account the current land use practices and settlement patterns. The zoning of the Baltic macroregion, conducted using the method proposed, shows that the effect of socioeconomic factors on the modern landscape development is heterogeneous. The regions in the north and northeast are affected to a lesser and those in the south and southwest to a greater degree. The latter regions' landscapes are strongly affected by the land use practices and the settlement patterns.

Keywords

Modern landscapes · Baltic macroregion · Landscape genesis · Effect of socioeconomic factors

E. A. Romanova (✉)
Institute of Environmental Management, Urban Development and Spatial Planning, Immanuel Kant Baltic Federal University, Kaliningrad, Russia
e-mail: alberta63@mail.ru

© Springer Nature Switzerland AG 2020
G. Fedorov et al. (eds.), *Baltic Region—The Region of Cooperation*,
Springer Proceedings in Earth and Environmental Sciences,
https://doi.org/10.1007/978-3-030-14519-4_25

Introduction

The extent of human influence on the populated territories' landscapes is such that the understanding of the development of modern landscapes is impossible without taking into account the socioeconomic factors at play. Therefore, modern landscape zoning requires a multiscale approach and it must discriminate among the landscape components of different levels, on the basis of both natural and socioeconomic parameters. The Baltic macroregion is an interesting case for a study into the landscape genesis. Although it is difficult to give an unambiguous definition of the Baltic region or to determine accurately its boundaries due to its considerable diversity, this concept is well-established and traditional and it is widely used in politics, economics, and socioeconomic geography (Klemeshev et al. 2017). Despite the heterogeneity of this territory in both natural and socioeconomic terms, its connection to the Baltic Sea influences all the spheres of life of local and regional communities and thereby facilitates the formation of the Baltic macroregion as a single geographic and geopolitical entity (Fedorov et al. 2017). In a narrow sense, the Baltic region includes the territories of Sweden, Finland, Estonia, Latvia, Lithuania, Denmark, a number of Russian regions (Leningrad, Pskov, Novgorod and Kaliningrad), several Polish voivodeships (Warmian-Masurian, Pomeranian and Western Pomeranian), as well as Germany's lands of Mecklenburg-Western Pomerania and Schleswig-Holstein. Despite its relatively small area, the region boasts a great variety of geographical phenomena, related to both its nature and society. The territory is also rich in borders and barriers of different kinds—national, ethnic, linguistic, and cultural, all of which affect the modern landscape environment.

Theoretical Background

At present, there are two main paradigms in studying the modern landscapes: the physiographic and the cultural ones. Russian landscape studies traditionally opt for physical geography. Within this paradigm, the natural factors that influence landscape formation are thoroughly scrutinised and the problem of human occupation, which is viewed as a factor of landscape transformation, is approached from the perspectives of rational land use and landscape geoecology—both these areas being closely related to physical geography (Isachenko 2008).

The western approach to landscape analysis, on the contrary, has been developing within the cultural paradigm. Since the works of Carl Ortwin Sauer, the landscape has been largely equated with its appearance (Sauer 1925; Keough 2013). Another area of research in the Western geographic literature focuses on the reconstruction of historical (and even archaeological) landscapes, which are associated with the land use and settlement patterns of the past. This field of study is known as "settlement archaeology" (Karro 2010; Rippon 2012). Interestingly, the definition of the term "landscape", offered by the European Landscape Convention, takes the human perception (i.e. the appearance) of landscapes into account and recognises the need for landscape assessment and the importance of landscape management (Jones 2007).

Different schools offer different views on landscape zoning. In Russia, landscape zoning has a long tradition with a focus on natural components. The most popular concept was developed by A. Isachenko. According to Isachenko (1991), the physiographical, or landscape, zoning includes zonal and azonal components. In the first case, the basic taxonomic unit is the landscape zone and, in the second, the 'physiographical' (or landscape) country (Isachenko 1991). At these levels, landscape zoning reflects the features of natural differentiation. At the regional level, especially in those areas where the natural environment is significantly affected by human activity, the "physiographical" approach to zoning is not effective. The structure and appearance of modern landscapes are affected by the socioeconomic factors and the main landscape-forming systems are the land use practices and settlement patterns.

The applied aspect of Russian landscape studies deals with landscape planning, which, however, has not been enshrined in law. Thus, in Russia, so landscape studies remain a theoretical discipline. Western landscape geography, on the contrary, is closely connected with spatial planning, which is a key tool for rational land use management and has a serious legislative framework. Land use management is closely associated with landscape zoning, since the landscape features (both natural and anthropogenic) determine the uses of territories and their significance for various aspects of environmental management. In Europe, spatial planning defines the land use patterns for each territory, including the rural and urban ones. In each country, spatial planning has its own specific features and depends on many factors. For example, in Bulgaria, agricultural lands are considered a national asset. Thus, changing the type of land use is allowed only in exceptional cases. Spatial zoning is mostly used to give a territory a protection status (Borisov 2015). The Nordic landscape management concept suggests a transition from the understanding of the term "landscape" in line with the definition from the European Landscape Convention ("an area, as perceived by people") to active landscape management (including relevant legislation and spatial management strategies) (Erikstad et al. 2015). The landscape management approach helps to draw up land-use guidelines for landowners and users (decision-makers). This approach requires concrete techniques for identifying and mapping "landscape zones". There are several distinct approaches to determining the main zoning criteria. According to one of them, landforms, ecosystems, and land use patterns can serve as the main zoning criteria. A different approach to mapping modern landscapes was suggested by the developers of the new European landscape classification, which uses digital data on climate, altitude, parent material and land use as determinant factors (Mücher et al. 2010).

Although landscape characteristics have been drawing the attention of both urban planners and landscape architects for a long time [at least since the emergence of the so-called "context theory" and the publication of I. McHarg's *Design with Nature* (McHarg 1969)], the problem of landscape zoning continues to arouse controversy among international researchers.

Methodology

This research aims at working out a model of integrated landscape zoning in the case of a single geographical region. In our opinion, traditional physiographical zoning, which is quite common for Russia, should serve as the basis for integrated landscape zoning, since natural factors prevail at the level of landscape zones and countries. At the same time, at the level of regions, it is vital to take into account the factors of anthropogenic differentiation, which depend entirely on the socioeconomic conditions.

Geographically, most of the countries of the Baltic macroregion have a humid continental climate. The only exceptions are the small areas in the north that belong to the subarctic zone. The Baltic macroregion is situated on the Baltic Shield and the Eastern European platform. The whole area belongs to the region of the last Quaternary glaciation—the Würmian. The region's landscape structure was shaped under the influence of the glacial and post-glacial natural processes. The topography of the Baltic macroregion is dominated by plains with a few relatively recent glacial forms—the terminal moraine uplands and ridges, kames, eskers and sandurs. There are the swampy ancient deltaic lowlands of the Vistula, Neman and Neva Rivers. The most elevated part of the region (the eastern slopes of the Scandinavian Mountains) lies in the north-west.

However, the appearance and structure of landscapes of the same natural genesis may vary because of the anthropogenic transformation of the natural environment. On the populated territories, the modern landscape environment depends directly on to what it has been transformed by human occupation, which in turn is affected by the socioeconomic factors. Landscape changes are thus a result of the land use and settlement patterns. Here, the Baltic macroregion is a good example. The anthropogenic impact on the soils and vegetation is not the same across the regions landscape zones because the territory of the Baltic macroregion is developed and populated rather unevenly. The natural factors of landscape formation prevail on the sparsely populated, almost intact territories. Such territories are typical for the northern part of the region. In other parts, particularly, in the South-West and the South, the densely populated territories betray the history of their anthropogenic transformation.

Thus, the process of landscape zoning on populated territories requires taking into account the factors of not only natural but also anthropogenic differentiation. To rank the territories depending on these factors, I used the indicator of the effect of socioeconomic factors on the landscape genesis (Romanova 2017). This synthetic measure shows the degree of anthropogenic differentiation of landscapes of any territorial unit. The indicator includes five components: the average density of the population, the forest and/or pasture coverage and agricultural use, the proportion of lands allocated to construction and roads, and the proportion of landscapes unaffected by human activity. Each component is measured on a scale from 1 to 5, thus the value of the indicator (the sum of the scores) ranges from 5 to 25. The higher the value of the indicator, the greater is the role of social factors in the landscape genesis.

This methodology does not contradict the existing methods of the landscape, historical-geographical, and cultural-landscape zoning, (Mil'kov and Gvozdetskii 1986; Andreev 2012; Isachenko 2013). On the contrary, it complements them by making it possible to identify the current and projected conditions of landscapes.

The calculation of the effect of the socioeconomic factors on landscape development in populated territories makes it possible to identify new landscape areas. This hierarchical level is confined to one landscape zone of one physiographical country or, in some cases, to one landscape. Therefore, the grid of the new landscape zoning is not used on its own but is superimposed on the existing grid of physiographical zoning. At this level, the patterns of land use and settlement acquire special significance for the landscape genesis and affect the appearance of the modern landscapes. These patterns may have a varying impact on landscape development—this explains why the territories characterised by the same natural features evolve into different landscape areas.

To establish the boundaries of the new landscape areas, one can use the gradient of the calculated synthetic indicator. The gradient of two contiguous administrative units scoring 6 and above suggests that the boundaries of the landscape area coincide with the administrative borders. The gradient of 2 and below suggests that the administrative units separated by administrative borders belong to a single landscape area.

The natural differentiation, which affects the genesis of modern landscapes, which are later exposed to either anthropogenic differentiation or natural evolution. While generally accepting the typological classification of landscapes proposed by A. Isachenko, I am convinced that there is a need for a new typological unit—the landscape subtype. This unit will reflect the degree of anthropogenic transformation of a landscape type. Having analysed the features of anthropogenic differentiation of the environment within each landscape type (for example, a glaciolacustrine plain), I identified the three main subtypes of landscapes. The first subtype comprises relatively natural landscapes. Their appearance is determined by the natural features (the topography, the quaternary deposits and bedrock, etc.). The structure of the composition stages is preserved. The other two subtypes are cultural and artificial landscapes. The former are changed and the latter created in the course of human occupation (Romanova 2017). In the landscape areas where the effect of the socioeconomic factors is insignificant, the relatively natural landscapes dominate. In the areas where the effect of these factors is strong, the prevalent landscapes are cultural and artificial.

Research Results and Discussion

To determine the key factors of landscape development on selected territories, I calculated the indicator of the effect of socioeconomic factors for the countries of the Baltic macroregion and their administrative units (NUTS 2 and 3 for the EU countries and regions for Russia). The areas associated with lower values of the

Fig. 1 The areas of the Baltic macroregion with the varying effect of the socioeconomic factors on the modern landscape genesis (A—natural factors prevail; B—the effect of the socioeconomic factors is middling; C—socioeconomic factors prevail; the figures indicate the values of the indicator)

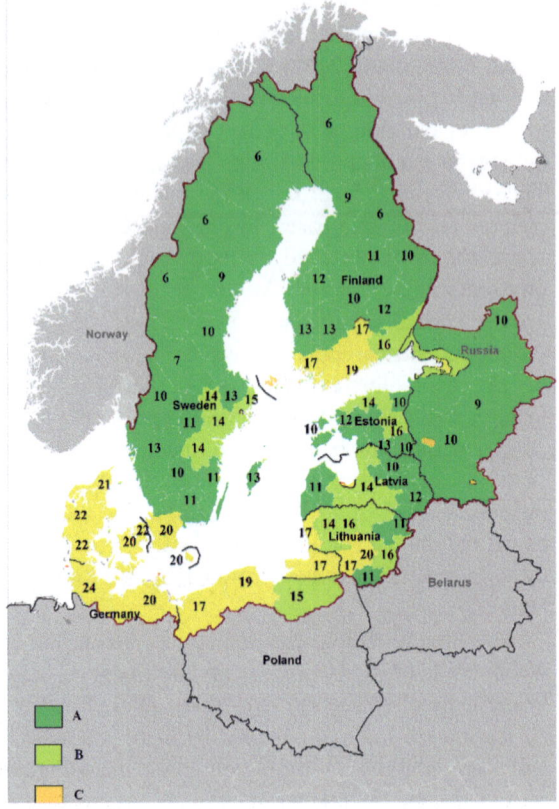

indicator, i.e. where the environment is almost unaffected by human occupation, are located in the north of the Baltic macroregion, in Latvia, Estonia, as well as in the Leningrad, Novgorod, and Pskov regions of Russia. The areas where the effect of the socioeconomic factors is the strongest are found in Denmark, Germany, South Sweden and the South-East Baltic. The settlement and land use patterns are the main factors of landscape genesis on these territories.

Over half of the territory of the Baltic macroregion is occupied by the areas that are strongly affected by natural factors. In the south of the region, the effect of the socioeconomic factors ranges from middling to strong. This holds true for the Kaliningrad region (Fig. 1).

Conclusions

The appearance and structure of the modern landscape are the results of both natural and anthropogenic influences, with the main factor of environment differentiation at the regional level being human occupation. The latter is manifested in the patterns of land use and settlement. The proposed synthetic indicator of the effect of socioeconomic factors on the modern landscape genesis can be used to differentiate among administrative and territorial units and to improve landscape zoning. To determine the boundaries between the modern landscape areas, the gradient of this indicator is of particular importance. The gradient shows whether the boundaries of a landscape area and the administrative borders coincide or not.

The values of the indicator of the effect of socioeconomic factors on the landscape genesis are represented vary across the Baltic macroregion. In the northern and northeastern parts, the effect is minimal, whereas the landscape genesis in the southern and southwestern areas is strongly affected by the socioeconomic factors. I suggest considering the settlement and land use patterns as the main factors of landscape genesis on these territories.

References

Andreev AA (2012) Opyt kul'turno-landshaftnogo raionirovaniya Rossii (Experience of cultural and landscape zoning of Russia). Pskovskii regionologicheskii zhurnal 13:12–25 (in Russ.)

Borisov B (2015) Spatial planning in regional planning of agricultural lands and rural areas. Bulg J Agric Sci 21:751–756

Erikstad L, Uttakleiv LA, Halvorsen R (2015) Characterisation and mapping of landscape types, a case study from Norway. Belgeo. https://doi.org/10.4000/belgeo.17412

Fedorov GM, Mikhailov AS, Kuznetsova TYu (2017) Vliyanie morya na razvitie ekonomiki i rasseleniya stran Baltiiskogo regiona (The influence of the sea on the development of the economy and the resettlement of the countries of the Baltic region). Baltic Reg 9(2):7–27 (in Russ.)

Isachenko AG (1991) Landshaftovedenie i fiziko-geograficheskoe raionirovanie: Uchebnik dlya vuzov (Landscape science and physical-geographical zoning: Textbook for high schools). Vysshaya shkola, Moscow (in Russ.)

Isachenko AG (2008) Landshaftnaya struktura Zemli, rasselenie, prirodopol'zovanie (Landscape structure of the Earth, settlement, nature use). St. SPbSU Publications, Petersburg (in Russ.)

Isachenko AG (2013) Printsipy istoriko-geograficheskogo raionirovaniya (na primere Severo-Zapada Evropeiskoi Rossii) (Principles of historical and geographical zoning (on the example of the North-West of European Russia)). Izvestiya RGO 145(1):3–20 (in Russ.)

Jones M (2007) The European landscape convention and the question of public participation. Landscape Res 32(5):613–633

Karro K (2010) Kodavere parish by Lake Peipus: the development of the cultural landscape during the Iron Age. Archaeologia Baltica 14:184–196

Keough S (2013) Cultural landscape. www.oxfordbibliographies.com. Accessed 7 Feb 2017

Klemeshev AP, Korneevets VS, Pal'movskii T, Studzhinitski T, Fedorov GM (2017) Podkhody k opredeleniyu ponyatiya «Baltiiskii region» (Approaches to the definition of the concept of the "Baltic region"). Baltic Reg 9(4):7–28 (in Russ.)

McHarg I (1969) Design with nature. Doubleday/Nature History Press, New York, NY, USA

Mil'kov FN, Gvozdetskii NA (1986) Fizicheskaya geografiya SSSR. Obshchii obzor. Evropeiskaya chast' SSSR (Physical Geography of the USSR. General review. European part of the USSR), vol 5. Vysshaya shkola, Moscow

Mücher CA, Kijn JA, Wascher DM, Joop HJ, Schaminée JHJ (2010) A new European Landscape Classification (LANMAP): a transparent, flexible and user-oriented methodology to distinguish landscapes. Ecol Ind 10:87–103

Rippon S (2012) Making sense of an historic landscape. Oxford University Press, UK

Romanova E (2017) Socioeconomic conditionality of the Baltic Macroregion landscape development and zoning. J Settl Spat Plann 8(2):131–137

Sauer CO (1925) The morphology of landscape. University of California, Publications in Geography, vol 2, issue 2, pp 19–53

The Main Directions in the Landscape-Ecological Planning of Urban Areas

Elena I. Golubeva and Tatiana O. Korol

Abstract

Modern conditions of urban growth and changes in the functions of urban areas dictate that the concept of urban planning ensures environmental safety, minimizes environmental risks, and provides for comfortable living and health of the population. The concept should promote positive emotional space and preserve cultural, historical, and spiritual heritage with the help of innovative technologies. It is essential to take the concepts of hazardous and safe urban environment into account when analyzing the properties of the territory that have direct impacts on the comfort of living. Solutions to complex problems related to urban environmental aspects should be based on models that incorporate information support systems aimed at urban social and economic development, landscape planning, elimination of the consequences of negative environmental impacts, the formation of the environmental framework, and improvement of the environmental situation in general.

Keywords

Urban areas · Urban environment · Landscape and environmental planning · Geographical factors · Environmental safety · Green building technologies

Introduction

A characteristic feature of any other phenomenon, including life in cities, especially large ones, is its duality. On the one hand, cities are appealing because they allow realizing human potential due to availability and diversity of job, education, professional growth and mobility opportunities, diverse contacts, and comfort

E. I. Golubeva (✉) · T. O. Korol
Department of Geography, Lomonosov Moscow State University, Moscow, Russia
e-mail: egolubeva@gmail.com

© Springer Nature Switzerland AG 2020
G. Fedorov et al. (eds.), *Baltic Region—The Region of Cooperation*,
Springer Proceedings in Earth and Environmental Sciences,
https://doi.org/10.1007/978-3-030-14519-4_26

economic, social, intellectual, transport, communal, and other conditions. On the other hand, negative aspects of life in cities are associated with high population density, crowdedness, unfavourable ecological situation, and stressful jobs and commutes.

In modern conditions of urban growth and changes in the function of urban areas, the concept of urban planning should ensure environmental safety, minimize environmental risks, provide comfortable living and emotional space, the health of the population, and preserve cultural, historical, and spiritual heritage with the help of innovative technologies. It is important to take the concepts of hazardous and safe urban environment into account when analyzing those properties of the territory that have a direct impact on the comfort of living.

Landscape-ecological planning of urban areas is a modern applied area of geographical science actively developing in Russia. It considers the diversity of the territorial organization of human economic activity taking into account environmental features of the territories and the conformity of the type of environmental management to planning and design of cultural landscapes (Smolitskaya et al. 2012).

Many Russian scientists have carried out assessments of living conditions of the population and the quality of life at various levels (from federal to local). For example, the Institute of Geography of the Russian Academy of Sciences has conducted the studies on the zoning of Russia's territory based on the natural conditions of life of the population (V. V. Pokshishevsky, E. A. Lopatina, O. R. Nazarevsky, L. A. Chubukov, Yu. N. Shvartseva, and others). In the late 1980s, A. N. Krenke and A. N. Zolotokrylin undertook regionalization studies dividing the territory of the North and East of Russia into districts and used quantitative and qualitative approaches to assess the population's living conditions. These studies have yielded generalized maps that primarily reflected specific climatic conditions.

Not only physical and economic geographers but also specialists in the field of human and medical geography have carried out the assessments of individual components of the environment and the analysis of their impact on people and their economic, recreational, etc. activities (A. P. Avtsyn, D. L. Armand, A. G. Voronov, A. G. Isachenko, B. B. Kochurov, G. M. Lappo, S. M. Malkhazova, Yu. N. Merinov, F. N. Mil'kov, B. B. Prokhorov, E. L. Reich, B. A. Revich, N. F. Reimers, V. S. Tikunov, Yu. P. Khrustalev, A. A. Shoshin, and others).

There have been assessments of the quality of urban environment conducted in various ways and using different methods based on both objective and subjective approaches. Common to most methods is the use of a more or less constant set of indicators in various combinations: characteristics of housing conditions, transport network, accessibility of services, environmental situation, public safety, and quality of landscaping (Bityukova 2003; Kurolap et al. 2010).

Assessment of the comfort of living and environmental safety of the urban environment requires greater attention to the analysis of geographical conditions, as well as the maximum integration of all environmental components. The integration of social factors with physical and environmental characteristics allows for a comprehensive assessment of the urban environment (Baraboshkina 2016).

From this perspective, the assessment of the comfort of living goes beyond the socioeconomic geography and becomes relevant for geoecological research that studies the problems of interaction between nature and society as a whole and specifically incorporates their spatiotemporal aspects (Ignatieva 2017). Solutions to complex problems related to urban environmental aspects should utilize models based on information support systems of management activities, which would facilitate social and economic development of cities, landscape planning, elimination of the consequences of negative environmental impacts, formation of the environmental framework, and the improvement of the environmental situation in cities in general (Vorobyova and Mogosova 2015).

The analysis of research conducted to date shows that it has targeted individual components of the urban environment or has considered isolated aspects of the issue. The authors suggest the strategy of a comprehensive approach to the landscape-environmental planning of urban territories.

Methods

A comprehensive study of the urban environment is impossible without the integrated approach based on modern databases, including remote sensing data of various levels of detail, and analyses that utilize geographic information systems (GIS); this provides for data integration and visualization and allows for a more effective monitoring of the state of the urban natural environment.

The authors have accumulated a large volume of data in the course of full-scale studies on the subject. The suggested approach involves the development of cartographic support that utilizes modeling methods and GIS technology for managing the quality of the urban environment in order to ensure environmental safety and comfortable living, and improve the quality of life of the population.

Results and Discussion

The material presented below describes the main features of the proposed approach to developing a comprehensive and multifactorial management system.

1. *Landscape and environmental planning* aim at the effective territorial organization of urban space, as well as at the overall optimization of urban planning. At the same time, consideration of cultural, historical, natural, aesthetic, and technological traditions within the framework of the active implementation of innovations should ensure the ecological safety of the territory by addressing the following aspects:

- *Economic or functional-production landscape planning* focused on minimizing the costs of economic activities at the regional and local levels. The leading role here belongs to engineering geography, natural-applied zoning (natural engineering zoning), and district planning.

- *Landscape-ecological planning* focused on the prevention or reduction of damage to nature from economic activity and the preservation or creation of favourable conditions for human life. Here the leading role belongs to geoecology and landscape ecology that study networks of specially protected natural areas within the environmental framework. The framework entails the creation of networks of protected natural areas with anthropogenic landscapes in between; the anthropogenic landscapes have centres for economic development. The foundation of the environmental framework is the network covering the most important territories and maintaining the landscape-ecological balance.

- *Aesthetic landscape planning* focused on the creation of an attractive and comfortable environment with landscape architecture and landscape-aesthetic design occupying the leading role. Currently, this is one of the most actively developing areas in the landscape and environmental planning of urban areas (Smolitskaya et al. 2012).

2. ***The analysis of the existing approaches to the assessment*** of the urban environment and current trends aims at the identification of ***the main directions of landscape-ecological urban planning*** (Fig. 1):

   - *Ensurance of environmental safety and minimization of environmental risks*: assessment of physical, chemical, biological, and aesthetic parameters of the environment; assessment and forecast of environmental risks.
   - *Optimization of the structure and functions of urban space*:

     - landscape-ecological planning of urban areas when changes in functional purpose occur;
     - protection of natural and cultural heritage;
     - organization of recreational areas with preservation of functions of nature protection and public health;
     - development of new types of settlements.

   - *Application of green technologies* in the planning and development of functional urban areas. All levels of design involve the use of landscape planning tools based on the ecological approach (sustainable design, "back-to-nature" and low impact approaches, etc.) (Sayanov 2013).

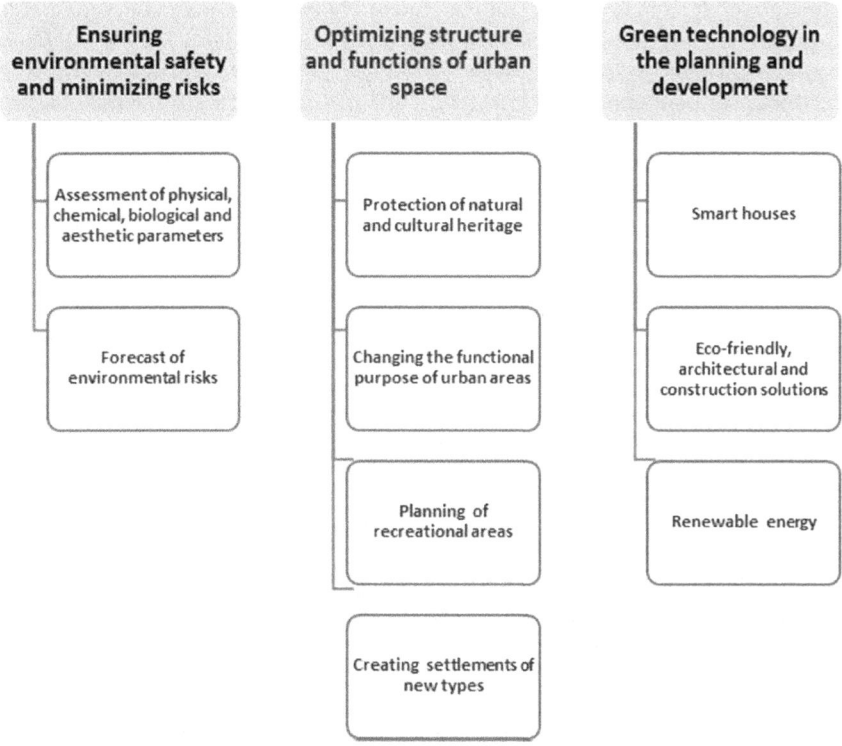

Fig. 1 The main directions in the landscape and environmental planning of urban areas

The features inherent in the suggested approach to the landscape-ecological planning of urban areas include:

– *Systems-based approach.* Systems analysis in landscape-ecological planning is used to study and plan natural-anthropogenic and cultural landscapes by constructing their ideal models. The models are oriented toward the identification of common harmonic components of nature.
– *Comprehensiveness and interdisciplinarity.* The paper addresses these aspects of the approach by the use of ecological-geographical, medico-ecological, landscape-planning, and cultural-historical methods in assessment and optimization of urban areas when their functional purpose changes; these methods comprise a single methodological platform, which ensures reaching the solution to the tasks and novelty of the research results.
– *Multiscale approach.* Large-scale level of research is most promising in urban planning. At the *regional-territorial level*, planning substantiations are developed for small areas, districts, individual settlements, industrial zones, and land plots, using 1:50,000—1:10,000 scale. *The local level* includes landscape and architectural studies and substantiation of planning projects for populated areas, industrial zones, and specially protected areas, detailed planning for the

construction of centres and residential and industrial areas of cities, and development of plans and land management projects (scale 1:25,000—1:2000). *The micro-territorial level* of landscape-ecological architecture and design is used for projects of development and design of settlements' centres, residential districts, industrial sites, urban and suburban park complexes, individual buildings, miniparks, and garden and cottage complexes (scale: 1:2000 and larger).

– *Computer modeling.* The widespread application of computer modeling allows solving various problems of green construction by recreating and analyzing the landscape and climatic conditions of local territories with the help of various specialized computer programs. The comprehensive approach to planning and design of urban areas at different territorial levels considers the following criteria: energy efficiency, lifecycle, thermal comfort, daylight, natural ventilation, solar shading, and building orientation with respect to the cardinal directions (Sayanov 2013). The results of the criteria's analysis are used to solve applied problems of landscape-ecological design and green construction:

– *Analysis of greenhouse gas emissions* allows obtaining environmental certification and achieving emission reduction; analyzing emissions with the maximal use of energy-efficient technologies, including renewable energy sources; and developing effective measures to limit the impact on the environment.

– *Analysis of solar shading* allows analyzing the results of solar ray tracing to assess the risk of undesirable shading of adjacent buildings and structures, exposure of translucent systems to daylight, and parameters of comfort in buildings and thermal loads on buildings.

– *Mathematical modeling of dynamic flows makes* it possible to conduct a detailed analysis of the state of air parameters and thermal flows inside and outside buildings. It is typically used to analyze the ventilation system, air conditioning, and heating; it allows predicting the spread of smoke or suspended particles in emergencies and man-made or natural disasters; it identifies critical zones for wind loading when planning residential districts and neighbourhoods.

– *Analysis of the potential, feasibility and economic impact of renewable energy sources* includes the assessment of the environmental component and financial feasibility of projects on the use of renewable energy sources to generate electricity and heat (photovoltaic systems, wind generators, solar collectors, biofuel systems, and geothermal and air heat pumps).

– *Accounting for the characteristics of the landscape and microclimate* (shading, wind loading, optimal orientation with respect to the cardinal directions, features of relief, and hydrogeological conditions) is crucial during various-size construction projects (Golubeva et al. 2015).

Conclusion

In order to formulate the concept of urban development from the standpoint of environmental safety and comfort of living, the article suggests a systems-based, comprehensive, and interdisciplinary approach with landscape planning of urban areas playing the crucial role. The basis for this approach is the laws and principles of landscape architecture, urban planning, ecology, sociology, psychology, landscape science, engineering geography, and many other fields of science.

To enhance the effectiveness of management measures in order to create ecologically safe, comfortable, and optimal urban environment in the third millennium, it is proposed to integrate the historical and geographical features of regions and innovative best available technologies of the XXI century.

Acknowledgements The work was supported in part by the RFBR project 15-05-01788 A.

References

Baraboshkina TA (2016) Integral'naya otsenka sostoyaniya osobo-okhranyaemykh territorii megapolisa (Integral assessment of the state of specially protected areas of the megalopolis). Interact Sci 10:2–5 (in Russ.)

Bityukova VR (2003) Ekologicheskaya situatsiya v Moskve v kontse XX veka (The ecological situation in Moscow at the end of the XX century). Izvestiya RAN. Ser. Geogr. 1:80–89 (in Russ.)

Golubeva EI, Korol TO, Sayanov AA (2015) Innovation in landscape design in urban areas. In: Congress proceedings of the 2nd world congress of the international federation of landscape architects "History of the Future", Saint-Petersburg, Russia, 10–12 June 2015. Peter the Great Saint-Petersburg State Polytechnic University Publishing, St. Petersburg, pp 633–637

Ignatieva M (2017) Lawn alternatives in Sweden—from theory to practice. Department of Urban and Rural Development, Swedish University of Agricultural Sciences, Uppsala

Kurolap SA, Nesterov YuA, Eprintsev AS (2010) Tipizatsiya territorii Voronezhskoi oblasti po urovnyu tekhnogennogo vozdeistviya na sredu obitaniya (Typification of the territory of the Voronezh Region in terms of the level of technogenic impact on the habitat). Vestnik VSU Ser Geogr Geoecol 1:5–11 (in Russ.)

Sayanov AA (2013) Landshaftno-ekologicheskii podkhod v planirovanii gorodskikh territorii (Landscape-ecological approach in the planning of urban areas). In: Zengin TYu (ed) Proceedings of the international scientific and practical conference rational nature management: traditions and innovations, Moscow, Moscow State University, 23–24 Nov 2012. Publishing House of Moscow University, Moscow, pp 301–302 (in Russ.)

Smolitskaya TA, Korol TO, Golubeva EI (2012) Gorodskoi kul'turnyi landshaft: traditsii i sovremennye tendentsii razvitiya (Urban cultural landscape: traditions and modern trends of development). Publishing House "LIBROKOM", Moscow (in Russ.)

Vorobyova TA, Mogosova NN (2015) Izuchenie sotsial'no-ekologicheskoi situatsii v gorodakh (A study of the social and environmental situation in cities). In: Proceedings of the international conference "InterCarto/InterGIS", Kuban State University, Krasnodar, 12–19 Nov 2015. Krasnodar, pp 553–560. https://doi.org/10.24057/2414-9179-2015-1-21-553-560 (in Russ.)

The Main Ways of Development in Agricultural Environmental Management in the Baltic Sea Region Countries (Historical and Geographical Analysis)

Olga L. Vinogradova ⓘ

Abstract

Agricultural nature management is one of the most ancient though dynamic branches of economy The author considers the development of the agricultural nature management systems of the Baltic Sea countries through the methods of historical and geographical studies and calculates the rates of change in the areas of cultivated land and the intensity of their use. The balanced regional model of nature management is based on the conjugate analysis of an array of statistical data for the period from 1909–13 to 2014–16. The method of process modeling for agricultural production development made it possible to identify key factors affecting the dynamics of the nature use of the Baltic sea region as a whole and its particular countries. The general dynamics tendency of agricultural nature management is seen as simultaneous compression of the areas of cultivated land and the growth of crop yields, while each of the Baltic sea countries has individual patterns of these processes. In accordance with the trajectories of agricultural development, the countries of the Baltic region are divided into three subregions. In the northern subregion, which includes Finland Sweden, and the Leningrad Region, agriculture has played an additional role in the nature management system over the past hundred years. In the southern subregion (Poland, Denmark, and Germany), agriculture is the leading type of nature management, but its role has been steadily falling throughout the period of research. In the southeastern part of the Baltic region (the Baltic countries and the Kaliningrad region) the agricultural nature management is gradually recovering after the 90s crisis and its role in the environmental management system of the subregion is increasing.

O. L. Vinogradova (✉)
Institute of Environmental Management, Urban Development and Spatial Planning,
Immanuel Kant Baltic Federal University, Kaliningrad, Russia
e-mail: OLVinogr69@mail.ru

© Springer Nature Switzerland AG 2020
G. Fedorov et al. (eds.), *Baltic Region—The Region of Cooperation*,
Springer Proceedings in Earth and Environmental Sciences,
https://doi.org/10.1007/978-3-030-14519-4_27

Keywords

Historical-geographical approach · The system of agricultural nature management · Regional models of nature management

Introduction

Agriculture and forestry, being the most land-intensive sectors of the economy,
reflect the global and regional trends in the development of the economies of
individual countries and regions, are sensitive to all external and internal factors.

A whole range of factors–like natural conditions, economic, political and
demographic situation, technological level–defines the main characteristics of
farming systems (its structure and spatial pattern, intensity) and determines the
trends of its development.

The main trend in the agriculture development in Europe is reduction in cultivated land area. One of the reasons for this process in Eastern Europe is the collapse
of the command economy system and institutional changes in the mid-1990s (Estel
et al. 2015; Horion et al. 2016). The decrease in the number of rural settlements also
leads to the abandonment of agricultural lands (Cartwright 2015; Stanaitis et al.
2011; Romanova et al. 2015). According to Jepsen et al. (2015) and Estel et al.
(2015), one of the trends in land management in the EU countries has been seen in
the conservation of landscapes, which reduces anthropogenic pressure on the
environment. The decrease in the area of agricultural land in Europe is also due to
increased crop yields, exports from other regions of the world (Lyury et al. 2010;
Levers et al. 2016). Cartwright (2015) notes that agricultural land is shrinking due
to growing demand for recreational purposes, expansion of nature protection zones
and construction of roads and urban settlements.

A number of factors influencing the agricultural nature management are global in
their nature, others determine the development of environmental management
systems for individual regions or countries (Jepsen et al. 2015; Levers et al. 2016;
Nikodemus et al. 2006; Vinogradova 2017).

The Baltic region includes countries with direct access to the Baltic Sea: Sweden, Finland, the Leningrad and Kaliningrad regions of the Russian Federation,
Estonia, Latvia, Lithuania, Poland, Germany and Denmark (Korneevets 2008). Due
to their geographic and geopolitical position, the countries of the Baltic region have
a number of common features conditioned by both natural conditions and the main
political and economic events.

A part of the region (the north of Sweden and most of Finland) is located within
the cold thermal belt, the rest of the territory lies in an average temperate belt
(Fedorov 2016; Romanova 1997). The cool thermal belt allows only local distribution of farming with a limited set of early grains and vegetables, fodder grasses,
the duration of the growing season in this territory does not exceed 120 days, the

total of effective temperatures varies from 1000 to 1600 °C. In the temperate zone, the sum of temperatures rises from 1600 °C in southern Finland and Sweden to 3000 °C in Germany and Denmark, and the vegetation period lasts up to 180 days. The entire territory of the region lies within the Atlantic Ocean influence, so the main challenge for agriculture is the widespread spread of wetlands, which has created complex and effective reclamation systems.

The general features of the evolution of agricultural land use are under the influence of political and economic events common to the whole Baltic region or individual countries: World Wars I and II, the revolution and disintegration of the Russian Empire, economic crises, technological revolutions in agriculture in the 1950s and 1970s, transition from the command to the market economy, changes in the agricultural products market.

Evolution Model of the System of Agricultural Nature Management in the Baltic Region

Complex historical and geographical studies allow us to identify the processes of formation and determine the specific ways of nature management development of each Baltic country and the Baltic region as a whole.

Statistical data from 1909–13 to 2015–16 included such indicators as the area of cultivated farmland (in% of the total area), the yield of cereals (in c/ha) (Muralov 1936; EUROSTAT 2016a, b; European Comission 2015; Federal Service of State Statistics 2017; Hansen 1922; OECD 2017; Hansen 1922; Demina 1959; Goskomstat of the USSR 1990; State Statistics Committee of the USSR 1991; League of Nations 1937).

For a proper comparison of agricultural land use condition in the countries and regions of the Baltic region, the statistical data from 1909–13 to 1946 were recalculated in accordance with modern boundaries. When identifying the development stages of agricultural land use within the region, economic and political developments that affect both the region as a whole and individual countries or regions were taken into account. A number of coefficients were accepted so that to assess the dynamics of the agricultural region intensity:

– the rate of change in the area of agricultural land in % a year from 1909–13 to 2014–15 and from 1994–95 to 2014–2015;
– the rate of increase in the yield of cereals in c/ ha per year for the same periods.

A calculation of indicators of changes in the intensity of agriculture over the past 20 years is necessary for a more accurate analysis of current trends, since this period is characterized by the most dynamic changes in the economies of a number of countries. An event-based historical-geographical approach as well as methods of historical and landscape research were chosen for the periodization of agricultural nature management dynamics (Nizovtsev and Erman 2015; Vampilova 2010).

A conjoint analysis of the factors and characteristics of each historical stage was used as a background for developing the balance and cartographic models of the agricultural nature management evolution in the Baltic Sea countries.

Evolution of Agricultural Nature Management in the Baltic Region

The countries and regions of the Baltic region are divided into three subregions for the entire study period: northern (Sweden, Finland and the Leningrad Region of the Russian Federation), southern (Poland, Germany and Denmark) and southeast (transitional), which includes Estonia, Latvia, Lithuania, and the Kaliningrad region. However, within these subregions, the dynamics of agricultural lands and changes in the intensity of nature management have their own features.

In Sweden and Finland, where the area of land suitable for agriculture is minimal, there has been a gradual but constant shrinkage of the range of cultivated land from 1934 to 36. The rate of decrease in the share of agricultural land for 100 years in these countries is minimal for the whole region (−0.01 and −0.02% per year, respectively).

The yield of grain crops over 100 years has increased more than three times. In Sweden, the growth was 40/ha (from 21.3 to 61.2) and its growth rate is one of the highest in the region (+0.40 c/ha per year); in Finland the rate of yield increase was not so sharp, which is apparently due to more severe natural conditions.

In the Leningrad Region, the agricultural development takes a different way: deep crises in the years after the revolution, the Great Patriotic War, and the post-war period, as well as the upsurge in 1934–36 and in 1975–78. From 1975 to 78 the area of agricultural cultivated land is constantly decreasing and amounts to 0.15% per year for a total of 100 years. For the Leningrad region, this situation is due not only to rather complex natural conditions, but also to the socio-economic role and position of the major metropolis and the industrial and transport specialization of the territory. The increase in yield over the past 20 years is almost the same as in Poland and Germany (+0.69 in the Leningrad region, +0.63 and +0.78 c/ha for the year in Poland and Germany), though it is the lowest in the region over the century (+0.26 c/ha per year).

In the Baltic States and the Kaliningrad region, from 1909–13 to 2014–16 the agricultural area was the largest in the region (from 61.6% in Latvia to 73.9% in Lithuania). In the subregion, there were similar recessions and upsurge in the years after World War II, and in East Prussia—after the First World War. After the collapse of the command economy in this subregion, a sharp decline in agricultural production occurred, although the decrease in the area of cultivated land in the Kaliningrad region began in 1978–80, while in Estonia, Latvia and Lithuania even earlier. A new upsurge, or rather, the restoration of agriculture began in 2003–04, in the Kaliningrad region in 2010–11. The growth rates over the past 20 years are the highest in the region—0.40–0.50% per year. But in general, over 100 years the area of agricultural land, as well as entirely in the region, is declining and falling faster than all other countries (down to −0.49% a year in Estonia). Currently, this southeastern subregion is intermediate between the northern and southern subregions considering the agricultural area. In general, for the whole period of the study the yield has grown at a moderate pace for the region, but in the last 20 years this growth has significantly accelerated (the maximum in the region is from +1.05 in Estonia to +1.25 c/ha per year in the Kaliningrad region).

In Poland, Germany and Denmark, throughout the studied period, the area of agricultural land was large (up to 81% in 1934–36 in Denmark). Despite the general decrease in the area of cultivated land, its rates are low—no higher than −0.15% per year. In Denmark, this process is slow and equable, while in Poland and Germany, there have been deep recessions in the years following World War I and World War II. In Poland, a new recession came after the collapse of the socialist bloc and in the latest 2–3 years, which may be due to the impact of mutual economic sanctions. The yield increase in these countries for 100 years is the highest in the whole region (up to +0.52 in Germany).

Conclusion

The dynamics of the agricultural nature management systems in the Baltic region during the period under study can be divided into several stages of decline and recovery, depending on the complex of political, economic and technological factors.

In each of the subregions, the development stages of agricultural nature management are more or less distinct: more clearly in the Baltic States, the Kaliningrad region and Germany, less obviously in Denmark and Sweden.

Three subregions can be distinguished due to some special impacts of a complex of factors and the evolution of agricultural nature management. The northern subregion includes Finland, Sweden, and the Leningrad Region of the Russian Federation. The area of agricultural land in Finland and Sweden did not exceed 10% and slowly declined over hundred years; in the Leningrad Region it fell from the mid-1990s. This subregion is characterized by a slow increase in crop yields. In the southern region (Germany, Denmark, Poland), agricultural lands occupy more than 50% of the territory, cultivated land shrinks slowly, though yields grow rapidly. The south-eastern subregion includes Lithuania, Latvia, Estonia and the Kaliningrad Region of the Russian Federation. After a sharp fall in the area of agricultural land, cultivated land has been undergoing some restoration since 2003–04 in the Baltic States and since 2010–11 in the Kaliningrad region, the fastest growing yield in the Baltic region.

In the Baltic States and the Kaliningrad region, high yields compensate for the decrease in the cultivated land area, so that gross harvest in 2014–16 was higher than in 1975–78, when the expansion of farmland was the highest in the post-war period.

References

Cartwright A (2015) Uncultivated agricultural land in Europe. Central European University. European Commission DG Enlargement Technical Assistance and Information Exchange (TAIEX). https://cps.ceu.edy/research/ua-land-europe. Accessed 16 Nov 2017

Demina VN (ed) (1959) Narodnoye khozyaystvo RSFSR v 1958 godu. Statisticheskiy yezhegodnik (The national economy of the RSFSR in 1958 Statistical Yearbook). Moscow State Statistical Publications (In Russ.)

Estel S, Kuemmerle T, Alcantara C, Levers Ch, Prishchepov AV, Hostern P (2015) Mapping farmland abandonment and recultivation across Europe using MODUS NDVI time series. Remote Sens Environ 163:312–325. https://doi.org/10.1016/j.rse.2015.03.028

European Comission (2015) Crop monitoring in Europe. MARS Bull 23(8). publication.jrc.ec.europa.eu/repository/bitstream/JRC98793/18-am-15-008-en-n/pdf. Accessed 3 Nov 2017

EUROSTAT (2016a) Agriculture, forestry and fishery statistics. KS-FK-16-001-EN.N.pdf. Accessed 26 Oct 2017

EUROSTAT (2016b) Land cover, land use and landscape. http://ec.europa.eu/eurostat/statistics-explained/index.php/Land_cover,_land_use_and_landscape. Accessed 10 Nov 2017

Federal Service of State Statistics (2017) Regiony Rossii. Sotsial'no-ekonomicheskiye pokazateli (Regions of Russia. Socio-economic indicators). http://www.gks.ru/bgd/regl/B15-14p. Accessed 30 Oct 2017 (In Russ.)

Fedorov GM (ed) (2016) Kaliningradskaya oblast'. Prirodnyye usloviya i resursy: ratsio-nal'noye ispol'zovaniye i okhrana (The Kaliningrad Region. Natural conditions and resources: rational use and protection). IKBFU Publications (In Russ.)

Goskomstat of the USSR (1990) Narodnoye khozyaystvo v Velikoy Otechestvennoy voyne 1941–1945 gg. (Statisticheskiy sbornik) (The national economy in the Great Patriotic War of 1941–1945. (Statistical compilation). Information and Publishing Center, Moscow (In Russ.)

Hansen J (1922) Einleitung zum Güteradzessbuck der Provinz Ostpreussen. www.ahnen-gesucht.de. Accessed 25 May 2016

Horion S, Prishchepov AV, Verbesselt J, de Beurs K, Tagesson T, Fensholt R (2016) Revealing turning points ecosystem functioning over the Northern Eurasian agricultural frontier. Glob Change Biol 22(8):2801–2817. https://doi.org/10.1111/gcb.13267

Jepsen MR, Kuemmerle T, Müller D, Erb K, Verburg PH, Haberl H, Vesterager JP, Andric M, Antrop M, Austrheim G (2015) Transitions in European land-management regimes between 1800 and 2010. Land Use Policy 49:53–64. https://doi.org/10.1016/j.landusepol.2015.07.003

Korneevets VS (2008) Ponyatiya «strany Baltiyskogo regiona» i «Baltiyskiy region» (The notions of "the countries of the Baltic region" and "the Baltic region"). Cosmopolis 2:68–77 (In Russ.)

League of Nations (1937) Statistical yearbook of the League of Nations 1936/1937, Geneva

Levers Ch, Butsic V, Verburg PH, Muller D, Kuemerle T (2016) Drivers of changes in agricultural intensity in Europe. Land Use Policy 58:380–393. https://doi.org/10.1016/j.landusepol.2016.08.013

Lyury DI, Goryachkin SV, Karavaeva NA, Denisenko EA, Nefedova TG (2010) Dinamika sel'skokhozyaystvennykh zemel' Rossii v XX veke i postagrarnoye vosstanovleniye rastitel'nosti i pochv (Dynamics of agricultural lands in Russia in the XX century and post-industrial restoration of vegetation and soils). GEOS, Moscow, pp 453–456 (In Russ.)

Muralov AI (ed) (1936) Sel'skoye khozyaystvo SSSR. Yezhegodnik 1935 g. (Agriculture of the USSR. Yearbook of 1935), Moscow (in Russ.)

Nikodemus O, Grine I, Peise Z, Rasa I (2006) Struktura landshaftov v Latvii: istoriya i trendy razvitiya (Landscape structure in Latvia: history and development trends). In: Dyakonov KN, Kasimov NS (eds) Landscape science: history, methods, regional studies and practice: Proceedings of the XI International Landscape Conference. Geography Faculty of Moscow State University, Moscow, pp 213–220 (In Russ.)

Nizovtsev VA, Erman IM (2015) Vozmozhnosti ispol'zovaniya istoriko-geograficheskikh materialov XVIII-XIX vekov dlya landshaftnykh issledovaniy (Possibilities of using historical and geographical materials of the XVIII-XIX centuries for landscape research). In: Bobrov EA (ed) Nature and society: in search of harmony. Proceedings of the International Scientific Conference, Smolensk, pp 104–115 (In Russ.)

OECD (2017) Agricultural land (indicator). https://data.oecd.org/agrland/agricultural_land.htm. Accessed 18 May 2018

Romanova EP (1997) Sovremennyye landshafty Yevropy (bez stran Vostochnoy Yevropy) (Modern landscapes of Europe (without the countries of Eastern Europe)). MGU Publication, Moscow (In Russ.)

Romanova EA, Vinogradova OL, Frizina IV (2015) Effekt szhatiya sotsi-al'no-ekonomicheskogo prostranstva v usloviyakh prigranich'ya (na primere SZFO) (The effect of squeezing the socio-economic space in the conditions of the border area (for example, the North-West Federal District). Baltic Reg 3(25):38–61 (In Russ.)

Stanaitis A, Stanaitis S, Subotkovichene R (2011) Dinamika sel'skogo naseleniya Litvy v XX veke (Dynamics of the rural population of Lithuania in the XX century). Pskov Regionological J 11:23–30 (In Russ.)

State Statistics Committee of the USSR (1991) Narodnoye khozyaystvo SSSR v 1990 g. Statisticheskiy sbornik (National economy of the USSR in 1990. Statistical compilation). Moscow (In Russ.)

Vampilova LB (2010) Teoriya regional'nogo istoriko-geograficheskogo analiza (Theory of Regional Historical and Geographical Analysis). Pskov Regionological J 10:129–140 (In Russ.)

Vinogradova OL (2017) Evolyutsiya modeley sel'skokhozyaystvennogo prirodopol'zovaniya stran Baltii i Kaliningradskoy oblasti (s 1890 po 2016 g.) (Evolution of models of agricultural nature management in the Baltic States and the Kaliningrad region (from 1890 to 2016). Bull Immanuel Kant Baltic Federal Univ named Ser Nat Med Sci 3:21–29 (In Russ.)

Spatial Development of the Petrodvortsovy District of St Petersburg: Primary Trends and Problems

Vasily L. Martynov⊙ and Irina E. Sazonova⊙

Abstract

The Petrodvortsovy district is one of the suburbs of St Petersburg. With 3% of the population of the city residing in this district covering about 8% of the city territory and containing over 20% of its historical and cultural monuments, it has a distinctive character. The research aims to describe transformational processes of the spatial structure of the Petrodvortsovy district as well as to systematize the data on main factors of its development. It also provides the analysis of the conditions determining the contemporary condition of the city and the view of possible changes. The methods employed to conduct the research are historical-geographical, comparative and analytical ones. The article identifies the prospects of development of primary spatial zones of the Petrodvortsovy district.

Keywords

Petrodvortsovy district · St petersburg · Spatial structure · Factors of its development

Introduction

A distinctive feature of Leningrad—St Petersburg is its suburbs, covering large areas and included into the city's territories. The area of St Petersburg according to its juridical borders is about 1400 km^2 with about 600 km^2 forming the city and about 800 km^2 being the suburbs. Until the end of the 1990s of the last century,

V. L. Martynov (✉) · I. E. Sazonova
The Herzen State Pedagogical University of Russia, Saint Petersburg, Russia
e-mail: martin-vas@yandex.ru

© Springer Nature Switzerland AG 2020 251
G. Fedorov et al. (eds.), *Baltic Region—The Region of Cooperation*,
Springer Proceedings in Earth and Environmental Sciences,
https://doi.org/10.1007/978-3-030-14519-4_28

most of the suburbs were not included in the list of the city territories. They had the status of "the territories of the Leningrad region subordinated by the administration of St Petersburg", and before, during the Soviet period, their status had been "the territories of the Leningrad region subordinated by the Leningrad soviet of the deputies. For this reason, all official reference books placed the suburbs among the territories of the Leningrad region. For example, the "Geography of Russia" encyclopedia, issued in 1998 (Gorkin 1998), stated: "Petrodvorets is a city in the Leningrad region".

By the end of the 1990s, the necessity for the final demarcation between the city and its suburbs became clear (Islyaev 2016). In this connection on 25 Dec. 1996 the Legislative Assembly of Saint Petersburg adopted the first in the city's history Act "About territorial organization of St Petersburg" (a new Act with the same name was adopted later in 2005), according to which the "subordinate territories" were included in St Petersburg officially. The border between the city and the Leningrad region has been determined once for all. Neither city nor regional government is interested in its changing. However, this border is somewhat bizarre: in some places, it runs along a street one side of which belongs to St Petersburg and another to the Leningrad region. However, this brought no cardinal changes in the life of the population of former "subordinate territories". Peterhof, Pushkin, Kolpino, Ses-troretsk and other towns and settlements of "subordinate territories" had never been connected with the Leningrad region. All these towns and settlements continue their existence safely in the borders of St Petersburg, keeping their faces, although G.M. Lappo assumed the reverse to happen (Lappo 2003). But the features and problems of development of suburban areas differ sharply from the ones of St Petersburg's districts – both central and marginal ones (Kryukova et al. 2016).

The Petrodvortsovy district is one of the "subordinate territories" that have been a part of St Petersburg since the end of 1990s. It accounts for approximately 8% of the city's territory, 3% of the population and 20.7% of historical and cultural heritage. By this indicator, The Petrodvortsovy district takes the second place in the city after the Tsentralny one (21.1%) (Lavrova 2013). Peterhof, Lomonosov and Strelna are so-called "palace suburbs", well known beyond St Petersburg. The most famous one is Peterhof with its fountains. The fountains are actually the only thing associated with Peterhof in mass consciousness.

The Development Before the Revolution 1917

The country residences in Peterhof, Strelna and Oranienbaum were created at the beginning of the XVIII century according to the plan of Peter the Great. The choice of location was very simple: Peter's Great St Petersburg was located on the road running along the coast of the Gulf of Finland from Swedish Nyen to Riga. As all other country residences, Peterhof was located on the distance of one-day horse ride from St Petersburg. The reason for choosing this particular location for "the capital of fountains" is that it lies on the side of the highest point of the Baltic Klint in the surroundings of St Petersburg—a limestone ledge stretching out along the coast of the Gulf of Finland and Ladoga Lake for several hundred kilometres. Here water streams flow to the gulf by gravity. This distinctive feature was used to create the fountains. However, natural streams were not enough, which is why there was the system of channels and reservoirs built and used to both supply the fountains and later provide Peterhof with water service. Thus, two factors determining the choice of the location for the residence of the emperor were anthropogenic (transport and geographic location) and natural one (relief peculiarity).

"The golden age" of the fountains of Peterhof fell on the reign of Elizabeth Petrovna, when the Peterhof Grand Palace, the Upper Garden and the Lower Park obtained their present shape.

Nicholas I gave the Lower Garden to public festivities. The Peterhof Grand Palace remained at the disposal of the Imperial Family. Although they did not use it as a residence, it stayed close for the public until 1917. Nicholas I created a new park, Alexandria, with the small Cottage Palace. There was an Orthodox church of St. Alexander Nevsky, known as "the Gothic Chapel" built in the park. His son Alexander II built one more palace, the Farm Palace. Nicholas II created the third palace, known as the Lower Palace.

The Lower Palace was the only Peterhof palace consciously destroyed during the years of the Soviet regime. In the early 1960s, "the Lower Palace" was blown up under the pretext of being in an emergency condition. In fact, its liquidation was associated with an attempt to "erase the memory" of the reign of the last emperor of Russia and of him. Now the Lower Palace is under reconstruction, but only its appearance is going to be reproduced. In the Soviet era, both the park and palaces were subjected to severe destruction. By the 1990s Alexandria has become practically "a forest park" with the remains of paths and drainage systems, a place to hold various "cultmass" activities. In the 10s of the 21st century, as a result of a great reconstruction, the park was restored

Nicholas I has also created the Colonist Park with the Olgin Pond. On the islands of this pond near the Tsaritsyn pavilion, there was "the Washington Oak" planted from the acorn, brought from the grave of the first US president.

Apparently, Nicholas I intended to give a "gothic" look not only to Alexandria but also to the city of Peterhof itself. During his reign, the Gothic buildings of the Imperial Stables and the Imperial Post were built; the only gothic building of the railway station in Russia—New Peterhof station—was projected and later, after the death of the emperor, built in the town.

Since the summer imperial residence needed protection, there were military units occupied mainly with guard duty in Peterhof. The army's Caspian regiment guarded the "external" frontiers, with the barracks located to the south of the Peterhof railway (partially used and partially neglected nowadays). The barracks of the Life Guards of the Ulan Regiment were located closer to the imperial palaces. The barracks of the Grenadier Regiment and the Life Guards of the Dragoon Regiment were near the palaces and parks in Old Peterhof. The completely closed complex of buildings of each regiment included barracks, stables, headquarters, houses of officers, various household buildings and regimental churches. With the abundance of "departmental" and manor churches, the first public church of Peterhof was built in the early years of the 20th century. It was the Peter and Paul Cathedral constructed in 1893–1905. The space between the imperial residences and garrisons of different regiments was built up in a chaotic way.

The roads connected Peterhof with other imperial residences, Tsarskoe Selo and Gatchina, and with the summer camps of the Guards in Krasnoe Selo as well.

Changes in Soviet Times

After the revolution of 1917 the decline of Peterhof, Strelna and Oranienbaum began. Life left the imperial palaces, although their doors were opened for the public, as they were turned into museums. The regiments guarding the peace of the emperors were disbanded, their barracks were empty. The mass outflow of the population began.

In the 1920s, the former imperial residences began to revive gradually together with Petrograd—Leningrad. An airfield of the Baltic Fleet was created at the former parade ground of military schools in Peterhof. Most of the churches were demolished, e.g. the square of Avrov is located at the site of the regimental temple of the Life Guards of the Ulan Regiment. Others were transformed into clubs, storehouses, etc. (the Church of the Caspian Regiment). Some of the palaces were converted into museums, others—into rest houses and boarding houses, etc., some were ruined. Soviet Peterhof was not a continuation of the pre-revolutionary Peterhof, it was a new city, standing in place of the former one. Actually, the same applies to Leningrad—St Petersburg (Martynov et al. 2008).

During the Great Patriotic War Peterhof and Strelna simply ceased their existence due to the fact that from September 1941 until January 1944 they were on the frontline at the German side. Germans took Peterhof easily in September 1941. Attempts to liberate the city were undertaken almost immediately; they cost many lives and had no result until the complete removal of the blockade in January 1944. Oranienbaum was on the frontline at the Soviet side, having undergone much less destruction. The palaces and parks of Lomonosov (Oranienbaum) are the only imperial royal residences preserved from imperial times, all the rest were destroyed and restored during the post-war decades.

It is a well-known fact that the restoration of the palace and park ensembles of Peterhof, first of all of the Lower Park, began at the end of the war (Petrov 2017). A less-known fact is that their accelerated reconstruction aimed at restoring Petrodvorets and the district.

A small lapidary factory that had existed in Peterhof since the beginning of the 18th century transformed into the Petrodvorets Watch Factory (1949). With new building constructed, it became one of the largest manufacturers of mechanical watches and other technical devices in the USSR. Much less famous but still significant is the company formerly known as "the 20th Armored Factory", now bearing the name of "61st Armored Repair Plant". In Strelna, the 55th Metal-working Plant of the Ministry of Defense was created. In Strelna, there was also the plant "Mashrybprom" built on the shores of the Gulf of Finland. A poultry farm later called "Krasnye Zori" was built at the intersection of Ropshinsky Highway and the railway In Petrodvorets. In Lomonosov, there also were industrial enter-prises and scientific institutions established. One of such institutions is "the Polar Marine Exploration Expedition".

The creation of new industrial enterprises made it possible to accelerate the pace of restoration and development of the Petrodvortsovy district. A new residential construction began.

Along with plants and factories, the importance of the city-forming enterprises arose from its military schools, e.g. the S. M. Kirov Combined Arms School (abolished later) and A. S. Popov Naval School of Radio Electronics located in Petrodvorets. In 1958, S. M. Kirov Combined Arms School was located in the former barracks of the Life Guards of the Ulan Regiment. Since 1947, the Naval Communication School has been located in the buildings of the former Military School of Emperor Alexander II. Since 1980 until now, the Military School of the Railway Troops, repeatedly changing its name, has been stationed in the former barracks of the Life Guards of the Dragoon Regiment. In 1946–1960, the Suvorov Military School of Frontier Troops was located in the barracks of the Caspian regiment.

The end of the 1960s saw the launch of the project of redeployment of A. A. Zhdanov Leningrad State University (current St Petersburg State University) to Petrodvorets. New buildings and hostels of Leningrad State University were built in the western part of Peterhof, so-called Old Peterhof. The construction here was carried out according to the initially approved and detailed plan (Polovcev 2013). The aim of the comprehensive development plan for the area to the south of the railway platform of Old Peterhof was to transfer the Leningrad University there, but it has never been and will never be completed.

During the 1980s and 1990s, military-scientific institutions were moving to Peterhof. The 24th Scientific Research Institute of the Ministry of Defense, founded as the Computing Center of the Navy, was located here. Today it is the Research Institute for Operational and Strategic Studies of Naval Construction, which is a part of the Naval Academy. The construction of a new building for the Institute began. However, in the early 1990s when the level of readiness reached about 90% it was frozen. For more than 10 years the building stood unfinished and abandoned, it acquired the nickname of "Pentagon" among the locals, then its construction was completed but, apparently, it became an elite residential building. In the 1990s, the military unit 45,707, a detachment of hydro-navists (researchers of ocean depths) of the Main Directorate of Deepwater Research of the Ministry of Defense, was quartered in Peterhof. It occupied some of the buildings of the abolished S. M. Kirov Combined Arms School.

Present-Day Condition

In the 1990s and the early years of the 21st century, the main industrial enterprise of the district, Petrodvorets watch factory, almost ceased functioning. Only some part of the premises of the former administrative building and a medical department remained. A shopping centre, "Raketa", occupied the rest of the premises. S. M. Kirov Combined Arms School was dissolved. The 55th Metalworking Plant of the Ministry of Defense in Strelna was destroyed. The activity of the "Krasnye Zori" poultry farm was discontinued. The disappearance of recreational facilities in the palace and park area is characteristic of the 21st century. "Petrodvorets" sanatorium, the basis of which was a complex of buildings of the Imperial Stables, and "Znamenka" boarding house in the estate with the same name were closed.

At the same time, the Konstantinovsky Palace in Strelna, which performs the functions of St Petersburg residence of the highest officials of the state, was restored. There was a road from Volkhonskoye highway with an overpass over the railway line laid to it.

Today, in the second decade of the 21st century, an active multistorey housing development is taking place in the areas of New Peterhof that were empty before. "Closed" residential single-storey communities are built as well, for example, between St. Petersburg Prospect and the railway platform "Krasnye Zori". However, this kind of development is typical for the northern suburbs of St Petersburg in general (Petri et al. 2012).

A powerful industrial hub is being created along Novye Zavody street, connecting Peterhof and Strelna. Within this unit, there is a new industrial zone "Maryino", a factory of "Bosch and Siemens" company ("BSH Bosch and Siemens"), for which a Karl Siemens street was created, as well as a special economic zone "Neudorf". A new port "Bronka" is under construction in Lomonosov.

Conclusion

In general, it is possible to distinguish between three main zones in the planning structure of the district.

The first is the area stretching between the coast of the Gulf of Finland and St Petersburg highway/avenue. This is the zone of "palaces and ruins". Within its limits, there are the main palace and park complexes, some of them restored in all their glory (the Lower Park with the Grand Palace, the Konstantinovsky Palace and all palaces of Alexandria), and some abandoned or semi-abandoned ("Znamenka" manor, the complex of the former barracks of the Horse-Grenadier Regiment). It is hardly possible to develop something here, only to preserve and to restore what already exists, as far as it is possible. The most of the territory of this zone is occupied by green plantations. This part of the district is the most interesting for tourists.

The second zone is between St Petersburg highway/avenue and the railway line of Peterhof. This is mainly the residential area of the 1950s and the 1960s with higher educational and scientific institutions, mainly military. Within the same zone, there is the Academic Gymnasium of the St Petersburg University, "Krasnye Zori" boarding school and other institutions of social infrastructure. Some historical parks are located here as well, they are used mostly for tourist purposes (Aleksandrovsky, Kolonistsky, Angliysky). However, the share of green plantations here is lower than in the first zone. Tourists come here, but only "while passing".

The third zone is located between the railway line and the border dividing St Petersburg and the Leningrad region. Today this part of the district is the most dynamically developing due to both residential and industrial construction. There are only a few parks here, the largest of which is Lugovoi Park. It stretches along the water channel feeding the fountains. This zone reaches its greatest power in Old Peterhof, but New Peterhof and Strelna are the most dynamic areas. For tourists, this territory is not interesting as a whole.

The main resource for the district development is its territory. The process of suburbanization determines the main prospects for its development. Comparison between the two shores of the Gulf of Finland shows that the northern shore (the Kurortny district) is traditionally more "prestigious" than the southern one (the Petrodvortsovy district). However, the northern shore has one "natural obstacle" for development: as the distance from the shore grows, its territory acquires a water-logged wooded character. As a result, 90% of its territory is not inhabited and it can hardly be populated. The Petrodvorets district is devoid of this disadvantage. Most of the territory of the district is "an ideal homogeneous plain". The advantage of the Kurortny district over the Petrodvorets one is its beaches, stretching along the northern shore of the Gulf of Finland as a continuous belt within St Petersburg, which practically do not exist on the southern coast. The Kurortny district is interesting mainly for recreational purposes, The Petrodvortsovy district – for touristic ones.

Restraining factors for the development of The Petrodvortsovy district are a continuation of its advantages. It is impossible to build up historical parks and/or demolish old buildings of any kind for the sake of new ones, regardless of their present condition. The second limiting factor is water supply and especially water drainage. The existing treatment facilities located on the shore of the Gulf of Finland cope with the load, but it is very difficult to increase it without expanding these structures, and there is nowhere to expand them. The third problem is the transport one. With the construction of the Ring Road (KAD), its importance for owners of personal vehicles fell, as they now have an opportunity to reach any district of St Petersburg quickly. Suburban trains are generally sufficient, but most of the region's population live at a significant distance from the railway stations and gets there by means of public transport. Buses spend a lot of time en route. The situation is facilitated by fixed-route taxis, but in "the peak hours" they are not enough. However, all these problems can be solved completely. Moreover, they are gradually being solved.

In general, The Petrodvortsovy district is developing steadily now. Its destiny, as it has always been, is connected with St Petersburg, repeating all the ups and downs of the "northern capital".

References

Gorkin A (ed) (1998) Geografiya Rossii (Geography of Russia). Bolshaya Rossijskaja Encycloprdia, Moscow (in Russ.)

Islyaev RA (2016) Administrativno-territorial'noe ustroistvo Sankt-peterburga: veroyatnye stsenarii preobrazovaniya v sisteme gosudarstvennogo upravleniya i mestnogo samoupravleniya (Administrative and territorial structure of St. Petersburg: probable scenarios of transformation in the system of public administration and local government). Ecomomika Severo-Zapada: problem I perspektivy razvitiia 1:112–122 (in Russ.)

Kryukova OV, Martynov VL, Sazonova IY, Polyakova SD (2016) Main spatial problems of St. Petersburg. Eur J Geogr 7(2):85–95

Lappo GM (2003) Eks-goroda Rossii (Ex-cities of Russia). Geografia. http://geo.1september.ru/index.php?year=2003&num=31. Accessed 15 January 2018 (in Russ.)

Lavrova TA (2013) Struktura i osobennosti razmeshcheniya ob'ektov istoricheskogo i kul'turnogo nasledfiya na territorii Sankt-peterburga (Structure and peculiarities of location of historical and cultural heritage objects on the territory of St. Petersburg). Peterburgskij Ekonomicheskij Zurnal 2:23–28 (in Russ.)

Martynov VL, Epikhin AA, Kononova GA (2008) Istoricheskaya geografiya Severo-Zapada (Historical geography of the North-West). RGPU im. A.I. Herzena, St. Petersburg (in Russ.)

Petri OV, Aksyonov KE, Krutikov SA (2012) Prigorodnye zakrytye zhilye kompleksy Sankt-peterburga: nachalo segregatsii ili smena obraza zhizni? (Suburban closed residential complexes of St. Petersburg: the beginning of segregation or a change in lifestyle?). Herald of St. Petersburg University. Series 7. Geology, Geography vol 1, pp 86–98 (in Russ.)

Petrov IV (2017) K voprosu o vosstanovlenii pervoi ocheredi fontanov Petrodvortsa v 1946 godu (To the question of restoring the first stage of the fountains of Petrodvorets in 1946). In: Proceedings of the "A maximus ad minima. Malye formy v istoricheskom landshafte" conference, St. Petersburg, 25–26 Apr 2016, pp 12–18 (in Russ.)

Polovcev IN (2013) Universitetskii kompleks v Petrodvortse—detishche akademika arkhitektury Igorya Fomina (University complex in Petrodvorets—the brainchild of academician of architecture Igor Fomin). Herald Civ. Eng. 5(40):52–57

Monitoring the Quality of Public Services at the Local Level—Polish Experience

Krzysztof Kopeć

Abstract

In several recent years, local government in Poland has shown a growing interest in monitoring, including monitoring of the quality of public services. It has also become of an increasing interest for researchers. Monitoring of the quality of public services can serve as a tool for raising the quality of life for the population, improving the functioning of individual self-government or administration institutions, and civilizing the social discourse. Monitoring of the quality of public services allows for a better understanding of the situation in local administrative units, informs whether changes in some aspects are for the better or worse, and is a useful tool in diagnosing the condition of local government/administrative units. The article presents the process of implementing and functioning of monitoring of the quality of public services and, above all, presents the experience gained from the implementation of this tool in Polish units of local government. The most important conclusions are the following: monitoring must not be an objective in itself; monitoring must address important issues; monitoring should require a small amount of work; monitoring is not intended for comparisons with other local government/administration units; monitoring must be resistant to tampering; no one should be afraid of monitoring; monitoring indicators should measure the effects rather than outlays; monitoring usually brings a lot of surprising results; the implemented system of monitoring is worth expanding and changing.

Keywords

Monitoring · Public service · Quality of public services · Local government · Administrative unit

K. Kopeć (✉)
Department of Regional Development Geography, University of Gdańsk, Gdańsk, Poland
e-mail: krzysztof.kopec@ug.edu.pl

© Springer Nature Switzerland AG 2020   259
G. Fedorov et al. (eds.), *Baltic Region—The Region of Cooperation*,
Springer Proceedings in Earth and Environmental Sciences,
https://doi.org/10.1007/978-3-030-14519-4_29

Introduction

The issue of monitoring is becoming more and more widespread in Poland. On the one hand, more local authorities have begun to express their interest in monitoring, and some of them have implemented the quality monitoring system for public services, on the other hand, monitoring, including monitoring of the quality of public services, is becoming a growing concern for researchers.

This article outlines the nature of monitoring, the quality of public services and the basic information about its implementation and operation. In addition, it presents experience gained from monitoring the quality of public services in Poland.

The Nature of Monitoring the Quality of Public Services

Monitoring is, in broad terms, generally understood as the process of regularly collecting and analyzing data (qualitative and quantitative) as well as the systematic assessment of certain phenomena carried out over a predetermined period, most often in strictly defined time intervals, and according to well-defined methods and a specific subject matter of the study (Czochański 2013).

The term 'monitoring' was rarely used in Poland until the 1980s. However, if used, it referred to two domains—education and environmental protection. Since then, the situation has changed considerably. Monitoring finds increasing application in other areas, among others, in public administration (Kopeć 2016). This is due to:

1. Growing awareness among local decision-makers of their role in the local development management, accompanied by a growing need to continually acquire more information, primarily on the needs of the residents, the state of gminas and poviats (Polish administrative units) and the effects of the actions taken (Kopeć 2016, 2017a; Michalski and Kopeć 2014, 2015).
2. Changes to the public management model. The Public Management model, which was bureaucratic, conformist, and reluctant to change has been abandoned. Public management has been transformed into a New Public Management model characterized by the introduction of managerial techniques to public sector management, increased productivity, and greater emphasis on the evaluation of effects than on processes, as well as decentralization. At present, the Public Governance model, which involves the meaningful participation of citizens in exercising authority, generated by the formation of civil society, is replacing this model. (see Bovaird and Löffler 2003, 2009; Hartley 2005).
3. Poland's accession to the European Union, which was crucial to accelerating the implementation of monitoring in the realities of the Polish public administration (Guidance document... 2014; Rokicki 2011; Zarządzanie i kontrola... 2007), also contributed to the development of this type of research and facilitated raising of funding for monitoring implementation (Michalski and Kopeć 2015).

Currently, monitoring is commonly regarded as a tool essential for proper development management and for effectively identifying the condition and changes occurring in the geographical space, and the spheres of human activity (Jacoby 2009).

Monitoring of the quality of public services focusing especially on local governments and administration is a particular type of monitoring. It can be used as a tool for (Michalski and Kopeć 2016):

– enhancing the quality of citizens' life in a given local government/administrative unit (Bugdol 2007; Kusterka-Jefmańska 2012, 2014; Rogala et al. 2012);
– improving the functioning of a particular self-government or administrative unit, i.e. improvement of management (Topolska-Ciuruś 2011);
– social discourse, which also contributes to the consolidation of democracy (Michalski 2014).

Monitoring the quality of public services to improve the quality of life of citizens should be regarded as superior to the other two (Michalski and Kopeć 2016).

Monitoring the quality of public services mainly fulfils three key functions in self-government/administrative units:

1. Monitoring allows for better understanding of the situation in the local government/administrative units, for finding the causes and consequences of processes taking place, and for recognizing the needs and aspirations of citizens;
2. It is similar to a thermometer indicating changes, which in some aspects are for the better, in some for the worse;
3. It is a tool useful in diagnosing the state of the local government/administrative unit, the development strategy or operational programs.

Implementation and Operation of Monitoring of the Quality of Public Services

Monitoring of the quality of public services in the local government/administrative unit is associated with the following characteristics:

1. The system is designed to be a low-cost solution, i.e. it does not require a lot of effort, time or money.
2. It involves a large number of people from multiple entities in a local government/administrative unit (e.g. gmina office, gmina social welfare centre, schools, etc.).
3. Monitoring is based on indicators that are not imposed. Those who implement monitoring activities decide which aspects and to what extent they will monitor.
4. Indicators should be measured annually.

The first step in introducing the monitoring of the quality of public services into a local government/administrative unit is to determine:

– what do we want to monitor?
– why?
– what for? (what do we want to achieve?)

The second step is the appointment of the monitoring coordinator, the person responsible for the monitoring system implementation.

It is important to implement monitoring following the steps below (Czerwińska et al. 2014; Michalski 2014; Michalski and Kopeć 2015):

1. Identify the needs, objectives and scope of monitoring;
2. Design the system;
3. Collect data and select indicators;
4. Analyze results and formulate preliminary applications;
5. Disseminate the results;
6. Debate and draw final conclusions;
7. Evaluate the monitoring and implementation of the conclusions.

It is important to repeat the major steps with each measurement, usually every year. This is because the implementation of monitoring is reasonable only under the assumption that it runs for a longer period. Monitoring in the short term does not have the potential to produce full effects.

The Polish Experience in the Implementation of Monitoring of the Quality of Public Services

The following data provide grounds for the presented conclusions:

– Standard Regional System of Monitoring the Quality of Public Services and Quality of Life created in the Gdańsk Institute for Market Economics[1]—later modified by the author in the field of monitoring the quality of public services;
– Experience gained during monitoring in Gdańsk and Słupsk—towns with poviat rights, Chojnice—municipal gmina, Czarna Woda and Siedliszcze—municipal and rural gminas, and the rural gminas of Stegna, Morzeszczyn, Puck and Mełgiew.

The whole system of monitoring the quality of public services consists of six thematic areas:

1. Administration;
2. Education and upbringing;

---

[1]It was presented in the publication "How to design and implement the system of monitoring the quality of public services and the quality of life? Implementation Manual" (Czerwińska et al. 2014).

3. Municipal economy;
4. Culture and leisure;
5. Help and care;
6. Environment.

In addition, for transparency purposes, each area divides into sub-areas for which the indicators are collected. The data for each indicator comprises of:

– data from records, e.g. office, school, social welfare centre;
– statistical data, primarily from the statistical office;
– surveys, mainly conducted among adult citizens of a local government/ administrative unit and among students in schools;
– self-assessment.

Basic recommendations and conclusions useful for the implementation of monitoring indicate that:

1. Monitoring cannot be an end in itself—it is only a tool for achieving other goals;
2. Monitoring must address important issues, as a result, indicators must measure important things;
3. Outlays (should be limited) and lack of visible short-term effects are the main reasons for obstructing monitoring;
4. Monitoring is not a tool for comparing self-governments/administrative units— monitoring is used to compare the current state of a local government/ administrative unit with its earlier state—one, two etc. years earlier;
5. Monitoring must be tamper resistant;
6. Monitoring should not cause fear in anyone (for example, the anxiety of administration officers may result in manipulation of the results or difficulties in the implementation process);
7. Monitoring indicators should measure effects and not the outlays (because the amount of money spent is not as important as the goals achieved);
8. Monitoring usually brings a number of surprising results—for example, bottom-up initiatives or detecting violence at school, which support further monitoring;
9. A monitoring system can be (and should be) complemented and changed if the character of a local government/administrative unit, its problems or objectives justify such steps.

Conclusion

This article only presents the basic information about monitoring the quality of public services and the general implementation experience in Poland. However, the subject matter is extensive. Polish academics address the question from various perspectives, e.g. as residents' satisfaction with the provided public services in the context of the quality of life (Wojtowicz et al. 2017), or as related to drug addiction monitoring and monitoring of other undesirable phenomena (Michalski 2010, 2015), or as monitoring conducted by particular self-governments—e.g. Słupsk (Szymańska 2016), or as monitoring conducted in specific administrative units— e.g. coastal gminas (Kopeć 2017b). This shows the popularity of monitoring the quality of public services in Poland among local governments and administrations as well as among researchers.

References

Bovaird T, Löffler E (2003) Evaluating the quality of public governance: indicators, models and methodologies. Int Rev Admin Sci 69(3):313–328

Bovaird T, Löffler E (2009) Understanding public management and governance. In: Bovaird T, Löffler E (eds) Public management and governance. Routledge, London-New York, pp 3–13

Bugdol M (2007) Rola administracji w kształtowaniu jakości życia. In: Skrzypek E (ed) Uwarunkowania jakości życia w społeczeństwie informacyjnym. Zakład Ekonomiki Jakości i Zarządzania Wiedzą UMCS, Lublin, pp 297–303

Czerwińska M, Gajdasz J, Hildebrandt A, Kopeć K, Kupc-Muszyńska B, Michalski T, Nowicki M, Susmarski P, Tarkowski M (2014) Jak zaprojektować i wdrożyć system monitoringu jakości usług publicznych i jakości życia? Podręcznik wdrażania, Instytut Badań nad Gospodarką Rynkową, Gdańsk. http://monitoring.ibngr.pl/wp-content/uploads/2015/02/Jak-zaprojektowac-i-wdrozyc-system.-Podrecznik-wdrazania.pdf. Accessed 25 Oct 2017

Czochański JT (2013) Monitoring rozwoju regionalnego. Aspekty metodologiczne i implementacyjne. Studia KPZK PAN, t. CXLIX, Polska Akademia Nauk, Komitet Przestrzennego Zagospodarowania Kraju, Warszawa

Guidance document on monitoring and evaluation. The programming period 2014–2020 (2014) European Commission. http://ec.europa.eu/regional_policy/sources/docoffic/2014/working/wd_2014_en.pdf. Accessed 26 Oct 2017

Hartley J (2005) Innovation in governance and public services: Past and present. Public Money Manage. 25(1):27–34

Jacoby Ch (2009) Monitoring und Evaluation von Stadt- und Regionalentwicklung. Einführung in Begriffswelt, rechtliche Anforderungen, fachliche Herausforderungen und ausgewählte Ansätze. In: Jacoby Ch (ed) Monitoring und Evaluation von Stadt- und Regionalentwicklung, Arbeitsmaterial der ARL, vol 350. Hannover, Akademie für Raumforschung und Landesplanung, pp 1–24

Kopeć K (2016) Aspekt transportowy w monitoringu jakości usług publicznych na szczeblu lokalnym. Autobusy—Technika. Eksploatacja, Systemy Transportowe 12:1799–1804

Kopeć K (2017a) Specyfika monitoringu aspektu transportowego w monitoringu jakości usług publicznych na szczeblu lokalnym. Autobusy—Technika, Eksploatacja, Systemy Transportowe 6:1709–1714

Kopeć K (2017b) Specyfika monitoringu jakości usług publicznych w gminach nadmorskich. In: Anisiewicz R, Połom M, Tarkowski M (eds) Rozwój regionalny i lokalny w perspektywie geograficznej, series: Regiony Nadmorskie. Wydawnictwo Bernardinum, Gdańsk-Pelplin, pp 106–121

Kusterka-Jefmańska M (2012) Pomiar jakości życia na poziomie lokalnym—wybrane doświadczenia europejskie i doświadczenia polskich samorządów. In: Borys T, Rogala P (eds) Orientacja na wyniki—modele, metody i dobre praktyki, Prace Naukowe Uniwersytetu Ekonomicznego we Wrocławiu, vol 264. Wydawnictwo Uniwersytetu Ekonomicznego we Wrocławiu, Wrocław, pp 230–239

Kusterka-Jefmańska M (2014) Jakość życia a jakość usług publicznych w praktyce badań na poziomie lokalnym. In: Potocki J, Ładysz J (eds) Gospodarka przestrzenna. Aktualne aspekty polityki społeczno-gospodarczej i przestrzennej, Prace Naukowe Uniwersytetu Ekonomicznego we Wrocławiu, vol 367. Wydawnictwo Uniwersytetu Ekonomicznego we Wrocławiu, Wrocław, pp 170–177

Michalski T (2010) Ewaluacja i monitoring narkomanii przez samorządy terytorialne. In: Ratajczak W, Stachowiak K (eds) Gospodarka przestrzenna społeczeństwu, vol 1. Bogucki Wydawnictwo Naukowe, Poznań, pp 207–217

Michalski T (2014) Propozycja systemu monitoringu jakości usług publicznych na szczeblu lokalnym. In: Ciok S, Janc K (eds) Współczesne wyzwania polityki regionalnej i gospodarki przestrzennej, vol 2. Series: Rozprawy Naukowe Instytutu Geografii i Rozwoju Regionalnego Uniwersytetu Wrocławskiego. Instytut Geografii i Rozwoju Regionalnego, Uniwersytet Wrocławski, Wrocław, pp 93–103

Michalski T (2015) Miejsce monitoringu przeciwdziałania narkomanii w całościowym programie monitoringu jakości usług publicznych. Serwis Informacyjny Narkomania 4(72):22–27

Michalski T, Kopeć K (2015) Monitoring jakości usług publicznych samorządów szczebla lokalnego narzędziem podnoszenia spójności społecznej. Studia i Materiały. Miscellanea Oeconomicae 4(1):139–150

Michalski T, Kopeć K (2016) Monitoring jakości usług publicznych narzędziem podnoszenia jakości życia na obszarach wiejskich. In: Biczkowski M, Rudnicki R (eds) Społeczno-ekonomiczny wymiar innowacyjności na obszarach wiejskich. Studia KPZK PAN, vol CLXXIII. Warszawa, pp 115–123

Michalski T, Kopeć K (2014) Monitoring yakosti publichnikh poslug organiv mistsevogo samovryaduvannya: pol's'kii dosvid (Monitoring the quality of public services of local governments: Polish experience). Publichne upravlinnya: teoriya ta praktika 4:17–24 [in Ukrainian]

Rogala P, Kusterka-Jefmańska M, Chojecka J, Koch-Mitka J (2012) Opis koncepcji, metodyki oraz narzędzi badań wskaźników jakości życia i wskaźników jakości usług publicznych dla jednostek samorządu terytorialnego na przykładzie Krakowa i Poznania. Kraków-Poznań. https://www.bip.krakow.pl/plik.php?zid=98139&wer=0&new=t&mode=shw. Accessed 24 Oct 2017

Rokicki B (2011) System monitorowania i ewaluacji polityki spójności w Polsce. Gospodarka Narodowa 3:87–104

Szymańska W (2016) Monitorowanie jakości usług publicznych problemem administracji samorządowej na przykładzie Słupska. Prace Komisji Geografii Przemysłu Polskiego Towarzystwa Geograficznego 30(2):171–185

Topolska-Ciuruś K (2011) Raport z pogłębionej diagnozy i analizy problemów, Śląski Związek Gmin i Powiatów. Katowice. https://benchmarking.silesia.org.pl/index.php?s=87. Accessed 24 Oct 2017

Wojtowicz A, Wojtowicz B, Nessel M (2017) Wykorzystanie narzędzi psychologicznych do pomiaru poziomu jakości życia mieszkańców terenów zurbanizowanych. Studia i Materiały. Miscellanea Oeconomicae 3(1):321–332

Zarządzanie i kontrola Funduszu Spójności. Ogólny podręcznik (2007) Ministerstwo Rozwoju Regionalneg, Warszawa

Innovation Security of Russian Borderland: The Case of Kaliningrad Region

Anna A. Mikhaylova⊙

Abstract

Innovation security is a new scientific agenda for human geography, the significance of which is growing with an increase in the need to form an innovation system in Russia in the context of unfavourable geopolitical conditions. The identification of territorial features of security is an important aspect of innovation security studies. The article aims to highlight the specifics of ensuring the innovative security of a border region. The study places emphasis on the influence of the border factor on the innovation security of a region in three contexts: geopolitical, institutional, and socio-economic. The proposed integrated approach to the system of innovation security is tested on the example of the Kaliningrad region—an exclave of the Russian Federation on the Baltic Sea. This region is selected due to its unique border position, high geopolitical significance and the actively developing innovation system. The article highlights the strengths and weaknesses of its regional innovation system in geopolitical, institutional, socio-economic aspects basing on the results of previously conducted empirical study on innovation security of the Kaliningrad region. This made it possible to draw a conclusion that the geopolitical context is of great importance for the innovation security of the Kaliningrad region. The article proposes a number of policy measures to strengthen the region's innovation system under unstable geopolitical conditions.

Keywords

Innovative security · Border region · Innovative development · Kaliningrad region

A. A. Mikhaylova (✉)
Institute of Environmental Management, Urban Development and Spatial Planning,
Immanuel Kant Baltic Federal University, Kaliningrad, Russia
e-mail: tikhonova.1989@mail.ru

© Springer Nature Switzerland AG 2020   267
G. Fedorov et al. (eds.), *Baltic Region—The Region of Cooperation*,
Springer Proceedings in Earth and Environmental Sciences,
https://doi.org/10.1007/978-3-030-14519-4_30

Introduction

The ongoing reorganization of the global geo-space architecture takes place in the context of two simultaneously developing opposite trends: the intensification of international socio-economic, geopolitical, cultural, historical, scientific, technological, military and other relations between one states and the complete or partial exclusion of others. This structural imbalance manifests in the formation of multi-scale nodes and peripheral zones that vary significantly in depth, breadth and frequency of integration functions and processes. The qualitative unevenness of the world economic system is a consequence of the trinity of interrelated effects of globalization-regionalization-polarization and competition for resources that primarily manifests in the strong socio-economic divergence of territories. In this regard, countries increasingly strive for sustainable development through the achievement of long-term competitiveness by creating strategic competitive advantages that cannot be reproduced elsewhere fully or in part. This involves the localization of competences in certain geographically and institutionally delineated boundaries. The competitiveness of a particular region is not least determined by the effectiveness of creating and using its innovative potential and the functioning of regional institutions take centre stage.

In the context of transition to the innovation economy model, based on knowledge as the vital resource and creativity as the key production factor, innovation security becomes a particularly important integral characteristic of national and regional security affecting the entire complex of relations and subsystems of the regional system (economic, social, scientific and technological, political and legal, ecological and geographical). It bears a direct relationship to competitiveness and sustainable development of a region. The concept of innovation security is relatively new, thus, it has no commonly accepted definition. The most general approach is to present innovation security as a component of economic security with an emphasis on the important role of innovation for sustainable economic development (Bagaryakov 2012; Golova 2014; Kormishkin and Saausheva 2013). However, taking into account an increasing importance of innovations for other spheres of society (social, political, environmental, etc.), innovation security has to be considered as a separate type of the national security system featuring its own subject, goals and objectives; objects and subjects of security; a set of special methods, mechanisms and tools.

A combination of internal and external factors determine the features of the regional system of innovation security. These factors include the level and nature of the innovation sector development, geographical and geopolitical features of governance, socio-economic conditions for the implementation of innovative processes, specificity of interests of socio-territorial communities as elements of a regional innovation system, a conflict of interests in the innovation sector at the inter- and intra-regional levels. Regional geo-economic position, i.e. the territorial location with respect to 'global knowledge pipelines' (Bathelt et al. 2004), proximity to knowledge and competence centres, involvement in transport networks, etc. have a great influence on regional innovation security. In this regard, it is of considerable interest to study the innovative security features of border regions.

Specificity of the Innovation Security System in the Border Area

Being geographically adjacent to the state border, this type of regions has a number of specific properties explained by the greater external influence of other states (comparing to that experienced by the internal regions) and by the fulfilment of special cross-border functions. These functions include assistance in ensuring border security, regulating external migration, monitoring the use of strategic resources, maintaining good relations with residents of the border areas of neighbouring states, creating conditions for the development of cross-border production relations, etc. The duality of the border contact-barrier function determines the differences in the interaction of the border regions of neighbouring states. According to Van der Velde and Martin (1997), these relations can be of the following types: alienated, coexisting, interdependent, and integrated. In combination with socio-economic, natural, climatic and other factors, this forms a unique profile of each border region and requires an individual approach to ensuring the security of its innovation system. However, there are general aspects of ensuring the innovative security of border regions distinguishing them from inland ones. Adopting the approach suggested by Morozova (2006), Prokop'yev and Kurilo (2016), it is necessary to consider the features of the innovation security system of the border region in three contexts: geopolitical, institutional, and socio-economic.

Geopolitical Context

Most models of regional economic development based on the innovation process focus on the development of interregional scientific, technological, and production cooperation, including that at the international level. In turn, the concept of security is focused on maintaining a high level of independence of the region from external actors in the most significant domains. In this case, the innovation security is a trade-off between openness as the most important requirement of an innovative economy and closeness as a condition for maintaining the security of the region.

In the geopolitical context, the innovation threats to border regions in comparison with inland regions are predominantly related to the following factors:

Migration: excessive systematic inflow of labor resources from neighbouring states with educational level and competencies not corresponding to the specialization and innovative profile of the region's economy, including low-skilled workers, refugees, illegal migrants who due to their professional and other qualities and other reasons are unable to integrate into the established innovative system of the region; significant migration outflow of young qualified specialists to more developed neighbouring countries;

Infrastructure: transport and energy blockade of the region by neighbouring countries; high dependence of regional innovative processes on the objects of the innovative infrastructure of neighbouring states or, on the contrary, complete infrastructure isolation;

Structural: strong dependence of the region's economy on foreign technologies; a low variety of trading partners located in neighbouring states; specialized business services rendered only by foreign organizations; foreign investments, including those directed at R&D sector;

Information: violation (including distortion) of new knowledge and information circulation between the border regions of neighbouring countries, leading to partial information isolation and the formation of an unreliable image;

Network: closed society, lack of desire for cooperation, experience exchange and knowledge sharing with neighbouring countries as a result of the implemented geopolitics.

The general geopolitical line is formed at the national level, in this regard, the innovation security strategy of a particular border region should be developed with respect to national priorities, which, however, may not fully coincide with regional interests. In general, the exploitation of the cooperation potential in the border area can be attributed to the vital need of the border region to ensure innovation security. The implementation of cross-border research and innovation projects, intensification of academic mobility, creation of joint innovation enterprises, establishment of professional contacts, etc. contribute to the innovative development of border regions. Thus, the creation of favourable conditions for cross-border cooperation should be regarded as a factor of innovation security.

At the same time, the use of this factor is associated with some threats connected with the nature and intensity of cooperation between the border regions. Firstly, if the interaction between the innovation process participants is not systematic, episodic and 'shallow' in nature, does not imply joint creation of new knowledge (joint R&D projects or full S&T cooperation) and, in essence, is reduced to simple trade relations, it does not lead to creation of conditions for innovation processes or their individual stages. Such cooperation is beneficial only for a more developed exporting region. It does not contribute to the formation of specific competitive competences in the importing border region. Therefore, in the case of the collapse of cross-border interactions, for example, due to the deterioration of the geopolitical situation, the main burden of breaking existing trade and production ties will lie on the less developed border region that depends on foreign technologies, components, specialized business services, etc.

Secondly, a unilateral outflow or an unbalanced exchange of ideas, knowledge, and technologies, when one of the border regions turns into a donor, and the other accumulates all the significant competitive competencies in the innovation sphere, is a significant threat to innovation security in the cross-border cooperation. As in the first case, the donor region forms a strong technological, infrastructural, personnel and other dependence compromising the security of its innovation system. In this regard, the best solution for ensuring the innovation security of each of the interacting border regions is the development of complementary competences, the best use of which is possible only with joint participation, excluding the possibility of a rapid change of partners. This model of cooperation presupposes equal interaction and a similar level of dependence, which helps to ensure a balance of relations between the parties.

Thirdly, the threat to innovation security of the border region is the targeted negative impact of the neighbouring country on its innovative system, including the hidden one. For example, if a foreign entity occupies a significant place in the funding structure of a particular field of research, it expands the opportunities for foreign influence on the research topics and the results obtained and facilitates knowledge drain. For the humanities and social sciences, significant external funding can present ideological and cultural threats.

Institutional Context

The institutional context of ensuring the innovative security of the border region is associated with the factors characterizing the functioning of formal and informal institutions involved in the innovation process. The study focuses on two important aspects of cross-border cooperation in the innovation sphere.

The first aspect largely relates to the activities of informal institutions and expresses itself as the level of trust between the cooperation actors in border regions. Trusting relationships significantly facilitate the process of knowledge and information exchange between participants in the innovation process, being a crucial cost-cutting factor (Camagni 1991; Wolfe 2002). The accumulation of social capital in the borderland and, ultimately, the formation of a transboundary innovation environment entail embeddedness of social relations in the cross-border context—a set of norms and rules of behavior that have emerged in the border regions due to economic, socio-cultural and other factors. Underestimation of the role of social capital in the development strategies of the border region is undoubtedly a negative factor for its innovation security. It is necessary to pay attention to the direct development of cross-border business cooperation, as well as to the creation of a borderland transport and business infrastructure, to the formation of a favourable customs and information space.

The second aspect includes political and legal bases for the cooperation of border regions in research, business, technological, educational and other spheres, including a system of laws and by-laws, regulation of activities of interstate and intergovernmental organizations, implementation of joint programs and initiatives. Factors that adversely affect the development of cross-border cooperation in the innovation sphere are the imperfection of legislation in the field of cross-border cooperation; significant differences in legislation regulating research and innovation; visa and customs barriers, etc.

Socio-Economic Context

The socio-economic context of innovation security in a border region is determined by the degree and nature of the impact of the border location factor on its economic and social sphere. In fact, these are the ongoing framework conditions for innovation activity. In this context, the threats to the innovation security of border regions in many ways overlap with the domestic ones: the high dependence of the economy on a narrow range of trade partners, including foreign ones; influx of low-skilled labor force in a volume significantly exceeding regional demand; stagnation in the real sector of the economy; sectoral focus on weak inter-industry links; weak transit function; low level and quality of life of the population, etc. Additional negative factors are political and economic instability, high bureaucratic, administrative, customs, tax barriers, lack of trust in public bodies, etc.

Innovation Security of the Kaliningrad Region

The Kaliningrad region is not just a border region of Russia, it is also an exclave region, and therefore the borderline factor here is particularly evident. According to the typology suggested by Malyavina (2004), the region belongs to the most significant border areas among Russian subjects along with the Leningrad region. Undoubtedly, it is of great interest as an object of research on the borderland features of innovation security. Based on the assessment results of the subjects of the North-West Federal District of the Russian Federation, the Kaliningrad region is classified as having an average level of innovation security (Mikhaylova 2017). The strengths of the regional innovation system include the high concentration of students and employed people with higher education, effective system of scientific personnel accession, digitisation of government's interactions with business and population (e-governance), a combination of fundamental and applied research in traditional and breakthrough areas, considerable inventive activity, experience in international economic activities, development of small business. The weaknesses are low innovative business activity, including lack of interest in financing research and development, training of personnel and the acquisition of patents and licenses for intellectual property use; technological backwardness; low awareness of environmental innovation; low productivity of labor and investment inflow; underfinancing of the research sector, including from the regional budget.

The geopolitical factor is of great importance to the innovation security of the Kaliningrad region. Being the Russian exclave in the Baltics, the region depends greatly on the EU-Russia relations. The most important domains that are subject to foreign policy influence are transport accessibility; mobility of the population and cargo; internal provision of the region with food and non-food products, raw materials and components; international scientific and technological cooperation within the framework of pan-European programs, EU framework programs, academic mobility partnership programs, joint multilateral initiatives and international projects in the Baltic macroregion.

Conclusion

The strong dependence of regional innovation activity on geopolitical factors pose a serious threat to innovation security of a region in the conditions of geopolitical instability. At the same time, a region has practically no mechanisms for influencing the geopolitical context. In this regard, a preventive measure for the Kaliningrad region may be the focus on improving the internal innovative milieu: increasing the efficiency of the regional innovation system, eliminating structural imbalances and gaps, finding and using internal reserves for a competitive breakthrough, and geographically differentiating economic ties and partners. In order to ensure innovation security at both national and regional levels, it is important to focus the implementation of innovative policies on its complexity (including not only economic but also social, environmental, cultural and other domains); interconnectedness enabling implementation of the regional innovation system approach to the management of regional innovation activities; adaptability based on the specifics and inalienable resources of the regions in determining their competitive advantages. For the Kaliningrad region, it is important to develop independent strategies for innovative development, specialized programs to support innovation, to expand innovation policy measures at the regional level, to increase regional budget expenditure on science, eliminate existing structural gaps in the innovation system, to strengthen innovative ties among economic entities within the region, improve local innovation infrastructure.

References

Bagaryakov AV (2012) Innovative security in the system of economic security of the region. Econ Region 2:302–305

Bathelt H, Malmberg A, Maskell P (2004) Clusters and knowledge: local buzz, global pipelines and the process of knowledge creation. Prog Hum Geogr 1(28):31–56

Camagni R (1991) Introduction: from the local « milieu » to innovation through cooperation networks. In: Camagni R (ed) Innovation networks: spatial perspectives. Belhaven Press, London, pp 1–9

Golova IM (2014) Substantiation of strategic priorities for ensuring innovation security of regional development. Econ Region 3:218–232

Kormishkin YeD, Sausheva OS (2013) Innovative security as a condition for the effective functioning of a regional innovation system. Reg Econ Theor Pract 34(313):2–8

Malyavina DA (2004) Economic-geographical approach to assessing the significance of the border regions of Russia. Department of Economic and Social Geography of Russia, Faculty of Geography, Moscow State University. http://ecoross.ru/old/malyav99.htm. Accessed 21 Aug 2017

Mikhaylova AA (2017) Evaluation of innovation security in Russian regions. Natl Interests Priorities Secur 4(13):711–724. https://doi.org/10.24891/ni.13.4.711

Morozova TV (2006) Karelian model of cross-border cooperation and the development of border local communities. Econ Sci Mod Russ 1:73–88

Prokop'yev YeA, Kurilo AYe (2016) Assessment of the impact of the border situation on the socio-economic development of the region (review of domestic literature). Pskov Regionological J 4(28):3–14

Van der Velde B, Martin R (1997) So many regions, so many borders. A behavioural approach in the analysis of border effects. In: Proceedings of the 37th European congress of the European regional science association, Rome, Italy, 26–29 Aug 1997

Wolfe DA (2002) Knowledge, learning and social capital in Ontario's ICT clusters. In: Proceedings of the annual meeting of the Canadian political science association University of Toronto, Toronto, Canada, 29–31 May 2002

The Migration-Related Risks and Balanced Labour Market Development in the Border Regions of Russia's Central and Northwestern Federal Districts

Anna V. Lialina⏺

Abstract

International labour migration (ILM) has a significant impact on the labour market development in the country of destination. This study aims at analysing the effect of ILM on the stability of labour markets in the Central (CFD) and Northwestern (NWFD) Federal Districts of Russia in terms of the markets' ability to resist possible economic risks and threats. I present a unique methodology for the index evaluation of a labour market's exposure to the basic economic migration risks and threats. This method includes an assessment of the indicators of revealed risks and threats, based on 13 criteria. The method was tested in the regions of the CFD and NWFD, which were subsequently classified into several types and clusters according to the degree of their labour markets' vulnerability to economic migration risks and threats. The applied method made it possible to identify the major effects of ILM on the labour markets. The border regions occupy a special place in the suggested typology, as they are characterised by a unique ability to attract and "absorb" foreign workers by integrating them into the regional labour markets. My analysis has shown that the border position of Saint Petersburg and the Leningrad and the Nenets Autonomous regions reinforces the negative effect of ILM. These regions show a heavy demand for foreign workers. In Saint Petersburg and the Leningrad regions, foreign workers are in great demand in construction, commerce, agriculture, manufacturing, and transport. In the Nenets autonomous regions, they are sought after in the mineral extraction industry. Other hallmarks of these regions are the increased labour market sensitivity to the risks of segmentation (by qualification, remuneration, profession, and industry) as a result of ILM and the emerging dependence on foreign individual entrepreneurship.

A. V. Lialina (✉)
Center of Modeling Socio-Economic Development of Region,
Immanuel Kant Baltic Federal University, Kaliningrad, Russia
e-mail: anuta-mazova@mail.ru

© Springer Nature Switzerland AG 2020   275
G. Fedorov et al. (eds.), *Baltic Region—The Region of Cooperation*,
Springer Proceedings in Earth and Environmental Sciences,
https://doi.org/10.1007/978-3-030-14519-4_31

Keywords

Labour market security · Risks and threats · External migration · International labour migration · Migration effects · Migration consequences for labour market · Demand for migrant workers

Introduction

The problem of assessing the impact of migration, and, particularly, its labour component on the development of a labour market, has been studies within various disciplines—economics, sociology and ethnocultural studies, geography, political science, jurisprudence, and others. Labour migration as an economic category was first addressed by economists in the 1960s in the classical and neoclassical models of economic growth. These include the Solow-Swan model of growth (1956) and the Ramsey growth model (1928). In one of the earliest works (Barro and Sala-i-Martin 2003), migration was considered as the main source of labour. The American scholars R. J. Barro and X. Sala-i-Martin demonstrated the negative impact of migration processes on economic growth, which contradicted the official statistics (Barro and Sala-i-Martin 2003). While the outflow of the population contributes to economic growth, the inflow, on the contrary, slows it down as the availability of labour increases and efficiency of the use of fixed assets decreases. However, the recent empirical studies (Neto et al. 2009; Lifshits 2013) have proven otherwise—the long-term effect of migration is usually either positive or insignificant.

The risks and threats to the economic development of the labour market have been studied by many Russian and international scholars within different theoretical frameworks. Some authors consider these problems from the perspective of the consequences of local residents and immigrants competing for jobs (with their professional heterogeneity taken into account). Others focus on the local resident's employment and remuneration (Angrist and Kugler 2001; Borjas 2003). These problems are examined in the context of the economic status of documented (Borjas 1987) and undocumented (Boswell and Straubhaar 2004) migrant workers and in terms of the economic efficiency of migrant workers' integration into the labour market (Roodenburg et al. 2003; Vorobieva 2005; Vasil'eva 2015; Solovieva 2016). Some authors analyse the "ethnic" specialization and segmentation of the labour market of the country of destination (Piore 1980; Chudinovskih et al. 2013; Ryazantsev 2016b) and address the dependence of the economy and its individual industries on the labour of migrants (Castles 2004; Tyuryukanova 2007; Martin 2010; Ivakhnyuk 2012). In a number of works, the risks and threats are viewed from the perspectives of immigrant integration into the labour market of the country of destination (Braun and Weber 2016) and from that of the labour migration mechanisms for in the conditions of the Soviet planned economy (Topilin 1975;

Khorev and Chapek 1978). The migration-related problems are studied in the context of the economic function of migration in the labour market (Khomra 1979; Rybakovsky 2003) and in terms of the impact of migration on the formation and functioning of the Russian model of the labour market (Topilin and Parfenceva 2006; Kashepov et al. 2008). Yet other authors examine the discrimination of foreign workers in the labour market (Vakulenko and Leukhin 2016), consider the forms, extent, and consequences of undocumented employment (Ryazantsev 2016a), and study the mechanisms for the generation of demand for migrant labour (Vakulenko and Leukhin 2015; Parfentseva and Ivanova 2011).

An analysis of these and other publications addressing the ILM-related risks and threats to national and regional economic security has made it possible to identify the key measures of the labour market development in the country of destination (Vitkovskaya and Panarin 2000; Ioncev 2003; Malakha 2005; Kulikov 2006; Glushenkov 2007; Gasparyan 2009; Gavrilova 2011; Dementyev 2012; Govorin et al. 2013; Kinasova 2013; Sidenko 2014; Savina 2015).

A. The increasing demand for migrant workers in the labour market.
B. The increasing demand for migrant workers in individual niches.
C. Loss of control and labour market criminalisation. Deteriorating labour conditions.
D. The emerging dependence of the labour market on foreign individual entrepreneurship.
E. The growing disparities among regional labour markets.
F. Labour market segmentation (by qualification, remuneration, profession, and industry).

Methodology

Index methods have been proven effective in evaluating economic security (Tseykovets 2015). In this article, I present a method for the index evaluation of Russian regions based on the degree of regional labour markets' exposure to migration-related risks and threats to the existing balance. This method relies on the quantitative analysis of the indicators listed above. Thirteen criteria, which meet the requirements of completeness, comparability, consistency, and representativeness, were selected to evaluate these indicators. These criteria include the migrant-worker-to-job ratio, the proportion of migrant labour in the workforce by industry, the regional localisation index of foreign workforce, the proportion of foreign entrepreneurs, the factor of labour market imbalance, the Duncan segregation index, and others. The official data from the Russian Federal State Statistics Service (Rosstat), the Ministry of Internal Affairs of the Russian Federation (before 2015, the Federal Migration Service), the Ministry of Labour and Social Protection, the Federal Service for Labour and Employment, and the Federal Tax Service served as the information sources for the research. The study employed the data on the companies' actual demand for migrant labour in 2010–2016.

At the first stage of the study, I selected and calculated the indicators necessary for my analysis. This required the evaluation of undocumented migration, allowing for the applicable adjustment factors. At the second stage, I set up the information and statistics database and calculated the necessary indicators. At the third stage, I performed the standardisation of indicators, using linear scaling. At the final stage, I calculated the economic index (EI), which reflects the exposure of regional labour markets to migration-related risks and threats and incorporates the individual indicator values according to formula 1, where indicator A is 15%, B 25%, C 15%, D 10%, E 20%, and F 15%. The EI values obtained served as the basis for the development of an interval typology and for cluster analysis based on the k-means algorithm. I used the IBM SPSS Statistics 23 software.

$$i_n = X_1 s_1 + X_1 s_2 + \cdots + X_n s_n \tag{1}$$

where i_n—economic index (EI),
X_n—the indicator,
s_n—the indicator's value in the economic index.

Results

Based on the level of a regional labour market's exposure to the risks and threats associated with economic migration, four types of regions were identified (Fig. 1; Table 1). A high level of labour market exposure to migration-related risks and threats was registered in two regions contiguous on the two capitals of Russia—the Moscow and the Leningrad regions. The four other regions are characterised by a moderate level of labour market exposure to migration-related risks and threats. These are the Tver and Tula regions in the Central Federal District and the Nenets autonomous region and St. Petersburg in the Northwestern Federal District. Note that, according to the calculated EI, almost half of the regional labour markets can be classified as relatively balanced and virtually unexposed to migration-related risks and threats. The overwhelming majority of the border regions of the Central and Northwestern Federal Districts of Russia (10 out of 13) fall into this category and only three of them (the Leningrad region, Saint Petersburg and the Nenets autonomous region) are at risk.

Clustering is carried out based on the similarities in the economic influence of international labour migration on the regional labour markets.

Cluster 1—The labour markets showing heavy demand for migrant workers (6 regions, including two border areas—St. Petersburg and Nenets AO). Cluster 1 is characterised by a high migrant-worker-to-job ratio, on the one hand, and the heavy demand for migrant workers, on the other. This holds true for the economy in general, individual industries, and self-employment. The distribution of

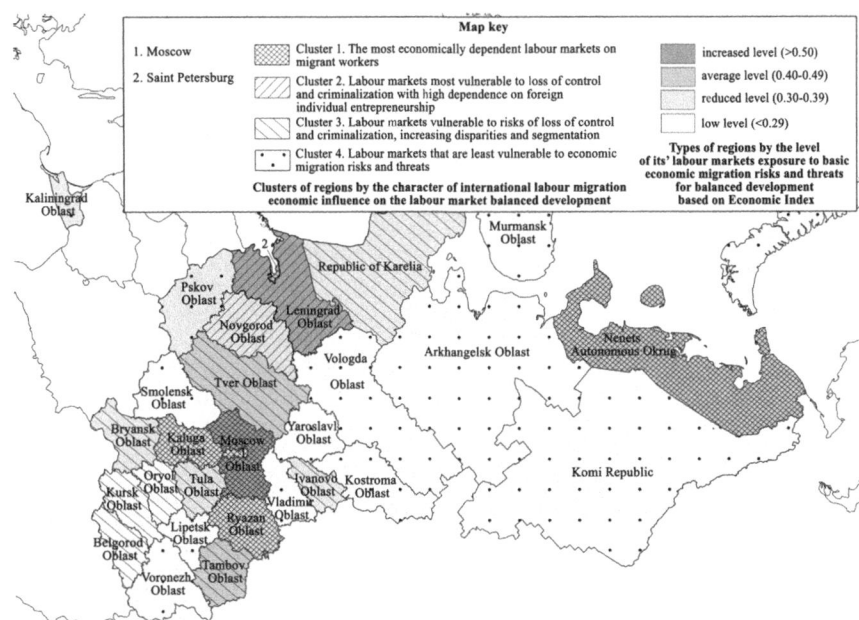

Fig. 1 The typology of the regions of the Central and Northwestern Federal Districts of Russia, based on the level of their labour markets exposure to the basic migration-related risks and threats. *Source* Calculated by the author

international migrant workers is closely related to the structure and demands of the regional labour markets. Today, labour migration mainly caters to the manufacturing industry, construction, agriculture, and commerce. The labour markets of Saint Petersburg, Moscow, and the Moscow region are showing strong demand for foreign workers in transport and communications. In the Nenets autonomous region, migrants are needed most in the mineral extraction industry. My research has shown that international migration often exacerbates the segmentation of the labour market by qualification, remuneration, profession, and industry. This is manifested in a) a significant discrepancy between the qualifications of migrant workers and of the people employed in the economy (most migrant workers performing manual labour); b) large wage differentials between migrant workers with different qualifications and the people employed in the economy; c) the high concentration of migrant workers in several industries, which hinders a balanced distribution of workers across the industries.

Cluster 2—The labour markets experiencing loss of control and criminalisation and having a considerable dependence on foreign entrepreneurship (the Leningrad and the Novgorod regions). Cluster 2 is characterised by a high correlation between the inflow of migrant workers and the development of undocumented employment, as well as a high proportion of foreign nationals and stateless persons among the self-employed and farmers. Note that, although classified as a cluster 2 region, the Leningrad region occupies the middling position. Based on many indicators, it

Table 1 The indicators comprising the economic index, calculated for the regions of the Central and Northwestern Federal Districts of Russia

		A. The increasing demand for migrant workers in the labour market	B. The increasing demand for migrant workers in individual niches	C. Loss of control and labour market criminalisation. Deteriorating labour conditions	D. The emerging dependence of labour market on foreign individual entrepreneurship	E. The growing disparities among regional labour markets	F. Labour market segmentation (by qualification, remuneration, profession, and industry).	Economic index
1.	Moscow region	1.000	0.691	0.000	0.914	0.235	0.594	**0.550**
2.	Leningrad region	0.485	0.390	1.000	0.891	0.232	0.461	**0.525**
3.	Nenets autonomous region	0.422	0.298	0.253	1.000	0.344	0.622	**0.438**
4.	Saint Petersburg	0.633	0.674	0.000	0.705	0.070	0.576	**0.434**
5.	Tver region	0.236	0.324	0.783	0.386	0.371	0.480	**0.419**
6.	Tambov region	0.079	0.118	0.781	0.000	0.824	0.515	**0.401**
7.	Ivanovo region	0.035	0.018	0.988	0.364	0.527	0.604	**0.390**
8.	Tula region	0.221	0.196	0.748	0.467	0.320	0.488	**0.378**
9.	Kaluga region	0.565	0.412	0.000	0.466	0.324	0.495	**0.373**
10.	Republic of Karelia	0.042	0.058	0.868	0.363	0.530	0.497	**0.368**
11.	Novgorod region	0.109	0.120	0.916	0.567	0.250	0.461	**0.360**
12.	Ryazan region	0.435	0.346	0.000	0.340	0.505	0.448	**0.354**
13.	Bryansk region	0.042	0.071	0.973	0.012	0.413	0.479	**0.326**
14.	Moscow	0.644	0.355	0.000	0.194	0.032	0.694	**0.315**
15.	Kaliningrad region	0.224	0.233	0.000	0.223	0.479	0.651	**0.308**
16.	Pskov region	0.084	0.154	0.000	0.652	0.578	0.510	**0.308**
17.	Smolensk region	0.122	0.120	0.000	0.847	0.479	0.464	**0.298**
18.	Republic of Komi	0.111	0.058	0.235	0.682	0.383	0.536	**0.292**
19.	Yaroslavl region	0.087	0.267	0.179	0.238	0.381	0.541	**0.288**
20.	Kursk region	0.037	0.028	0.483	0.219	0.436	0.490	**0.268**
21.	Vologda region	0.069	0.097	0.383	0.339	0.364	0.410	**0.260**
22.	Oryol region	0.088	0.064	0.354	0.021	0.497	0.476	**0.255**
23.	Vladimir region	0.180	0.154	0.000	0.276	0.368	0.465	**0.237**
24.	Belgorod region	0.097	0.081	0.477	0.072	0.251	0.419	**0.227**
25.	Murmansk region	0.027	0.037	0.000	0.637	0.355	0.468	**0.218**
26.	Kostroma region	0.070	0.130	0.000	0.207	0.421	0.459	**0.217**
27.	Arkhangelsk region	0.002	0.013	0.000	0.233	0.509	0.557	**0.212**
28.	Lipetsk region	0.093	0.115	0.000	0.357	0.282	0.452	**0.203**
29.	Voronezh region	0.082	0.063	0.000	0.109	0.386	0.539	**0.197**

Note The border areas are highlighted
Source Calculated by the author

gravitates towards subcluster 1. All the region's industries, with the exception of mineral extraction, show a heavy demand for migrant workers.

Cluster 3—The labour markets experiencing loss of control and criminalisation and facing a considerable risk of growing imbalances and segmentation. This cluster comprises nine regions (including the border regions of Belgorod, Bryansk and Kursk and the Republic of Karelia). They are facing the following risks: an increase in undocumented employment, the growing supply and demand imbalance as regards migrant labour supply, and a deepening segmentation by qualification, remuneration, profession, and industry. The Tver and Tula regions obviously belong here, since they are facing the imbalance threat as a result of their manufacturing industry showing a strong demand for migrant labour.

Cluster 4—Labour markets that are least exposed to migration-related risks and threats. Cluster 4 is the most numerous. It includes twelve regions, including six border territories (the Pskov, Murmansk, Smolensk, Voronezh, Arkhangelsk, and Kaliningrad regions). For example, such border areas as the Voronezh, Smolensk, Arkhangelsk, Kaliningrad, and Pskov regions are not facing considerable risks relating to imbalances and segmentation. The Kaliningrad region (as well as some of the other regions in this cluster) is exposed to the risk of a heavy demand for migrant labour in individual industries (primarily, construction and manufacturing).

Conclusion

The proposed method for assessing the level of labour market exposure to migration-related risks and threats makes it possible to classify the regions of the Central and Northwestern Federal Districts of Russia and group them into several clusters. My findings can be instrumental in producing recommendations for the prevention and mitigation of risks and threats in view of the region's type and cluster. My analysis has shown that the most important risks are those of growing imbalances and deepening labour market segmentation, as well as those of loss of control and criminalisation caused by the inflow of the irregular labour migrants and their undocumented employment. The metropolitan regions, the contiguous and intensively developing Kaluga and Ryazan regions, and the mineral-extracting autonomous region are highly exposed to these risks. The border position can have a varying effect on the region. In Saint Petersburg, the Leningrad region, and the Nenets autonomous region, it aggravates the negative impact of labour migration, whereas, on the other territories, its role is insignificant.

Acknowledgements The author expresses her gratitude to E. Vakulenko, R. Leukhin, and A. Sirotinsky for their assistance in the preparation of this study.

References

Angrist JD, Kugler AD (2001) Protective or counter-productive? European Labour Market Institutions and the Effect of Immigrants on EU Natives. NBER Working paper 8660. http://www.nber.org/papers/w8660.pdf. Accessed 15 Mar 2017
Barro RJ, Sala-i-Martin X (2003) Economic growth, 2nd edn. The MIT Press, Cambridge, Mass
Borjas GJ (1987) Self-selection and the earnings of immigrants. Am Econ Rev 77(4):531–553
Borjas GJ (2003) The labour demand curve is downward sloping: reexamining the impact of immigration on the labour market. Q J Econ 118(4):1335–1374

Boswell C, Straubhaar T (2004) The illegal employment of foreign workers: an overview. Intereconomics 39(1):4–7

Braun S, Weber H (2016) How do regional labour markets adjust to immigration? A dynamic analysis for post-war Germany. Discussion paper Deutsche Bundesbank, Frankfurt am Main

Castles S (2004) Why migration policies fail. Ethnic Racial Stud 27(2):205–227

Chudinovskih OS, Denisenko MB, Mkrtchyan NV (2013) Vremennye trudovye migranty v Rossii (Temporary labour migrants in Russia). Demoscope Weekly. http://demoscope.ru/weekly/ 2013/0579/demoscope579.pdf. Accessed 16 Sept 2017 (in Russ.)

Dementyev NV (2012) Sovershenstvovanie sistemy obespecheniya ekonomicheskoi bezopasnosti Rossii v sfere vneshnei trudovoi migratsii (Improvement of the system of ensuring Russia's economic security in the sphere of external labour migration). Dissertation, Tambov State University named after G.R. Derzhavin (in Russ.)

Gasparyan VV (2009) Otsenka vliyaniya migratsionnykh protsessov na ekonomicheskuyu bezopasnost' Rossii i ee regionov (Assessment of the impact of migration processes on the economic security of Russia and its regions). Dissertation, North Caucasus Federal University (in Russ.)

Gavrilova TM (2011) Vliyanie migracionnyh processov na ehkonomicheskuyu bezopasnost' Rossii (Influence of migration processes on the economic security of Russia). Nat Interests Priorities Secur 27(120):67–74 (in Russ.)

Glushenkov AM (2007) Protivodeistvie tenevym migratsionnym protsessam v sisteme obespecheniya ekonomicheskoi bezopasnosti Rossiiskoi Federatsii (Counteraction to shadow migration processes in the system of economic security of the Russian Federation). Dissertation, Moscow University of the Ministry of Internal Affairs of the Russian Federation named after V.Y. Kikot (in Russ.)

Govorin A, Kaziakhmedov G, Kinasova E (2013) Osobennosti vliyaniya immigratsii na ekonomicheskuyu bezopasnost' Rossii (Features of immigration influence on economic security of Russia). Vestnik Instituta Ekonomiki Rossiyskoy akademii nauk 1:120–129 (in Russ.)

Ioncev VA (ed) (2003) Migratsiya i natsional'naya bezopasnost'. Nauchnaya seriya: Mezhdunarodnaya migratsiya naseleniya: Rossiya i sovremennyi mir (Migration and national security. Scientific Series: International Migration of Population: Russia and the Modern World), vol 11. Makspress, Moscow (in Russ.)

Ivakhnyuk IV (2012) Migracionnaya vzaimozavisimost'. CHast' II «Evrazijskaya migracionnaya sistema: ot ehkonomicheskogo pragmatizma k vozrozhdeniyu civilizacionnogo edinstva» (Migratory interdependence. Part II "The Eurasian migration system: from economic pragmatism to the revival of civilizational unity"). http://www.gumilev-center.az/ migracionnaya-vzaimozavisimost. Accessed 09 February 2017 (in Russ.)

Kashepov AV, Sulakhshin SS, Malchinov AS (2008) Rynok truda: problemy i resheniya (Labour market: problems and solutions). Nauchnyi ekspert, Moscow (in Russ.)

Khomra AU (1979) Migratsiya naseleniya: voprosy teorii i metodiki issledovaniya (Population shift: questions of the theory and research methodology). In: Steshenko VS (ed) Academy of Sciences of the Ucrainian SSR, Institute of Economics. Naukova dumka Publ., Kiev (in Russ.)

Khorev BS, Chapek VN (1978) Problemy izucheniya migracij naseleniya (Problems of studying migration of population). Mysl', Moscow (in Russ.)

Kinasova ED (2013) Immigratsiya i ekonomicheskaya bezopasnost' Rossii (Immigration and Economic Security of Russia). Dissertation, Moscow University of the Ministry of Internal Affairs of the Russian Federation named after V.Y. Kikot (in Russ.)

Kulikov VA (2006) Nelegal'naya migratsiya i ee vliyanie na ekonomicheskuyu bezopasnost' strany: organizatsionno-ekonomicheskie aspekty (Illegal migration and its impact on the economic security of the country: organizational and economic aspects). Dissertation, Management Academy of the Ministry of the Interior of the Russian Federation (in Russ.)

Lifshits M (2013) The influence of migration and natural reproduction of labour force upon economic growth in the countries of the world. Appl Econometrics 31(3):32–51

Malakha IA (2005) K voprosu o vliyanii migracionnyh processov na ehkonomicheskuyu bezopasnost' (On the issue of the impact of migration processes on economic security). In: Ispol'zovanie inostrannoj rabochej sily: problemy i perspektivy. IIEPS RAS, Moscow, pp 38–43 (in Russ.)

Martin P (2010) A need for migrant labour? UK-US comparisons. Who needs migrant workers? In: Martin R, Anderson, B (eds) Labour shortages, immigration, and public policy. Oxford University Press, Oxford, pp 295–323

Neto JPJ, Claeyssen JCR, Ritelli D, Scarpello GM (2009) Migration in a Solow Growth model. SSRN. http://papers.ssrn.com/sol3/papers.cfm?abstract_id=1578565. Accessed 16 Aug 2017

Parfentseva OA, Ivanova NP (2011) The assessment of labour demand of Russian economy in view of the prospective regional socio-economic development. IKBFU's Vestnik Ser. Humanit Soc Sci 3:41–47 (in Russ.)

Piore MJ (1980) Birds of passage: migrant labour and industrial societies. Cambridge University Press, New York

Roodenburg HJ, Euwals RW, ter Rele HJM (2003) Immigration and the Dutch Economy. CPB Special Publications series, vol 47. CPB Netherlands Bureau for Economic Policy Analysis, Hague

Ryazantsev SV (2016a) Vklad trudovoi migratsii v ekonomiku Rossii: metody otsenki i rezul'taty (Contribution of labour migration to the economy of Russia: evaluation methods and results). Gumanitarnye nauki. Vestnik finansovogo universiteta 2(22):16–28 (in Russ.)

Ryazantsev SV (2016b) Rol' trudovoj migracii v razvitie ehkonomiki Rossijskoj Federacii (The role of labour migration in the development of the Russian Federation economy). http://www.unescap.org/sites/default/files/1%20Role%20of%20Labour%20Rus%20report%20v3-1-E.pdf. Accessed 11 Feb 2017 (in Russ.)

Rybakovsky LL (2003) Migratsii naseleniya (voprosy teorii) (Migration of population (theory issues). The ISPR RAS, Moscow (in Russ.)

Savina SE (2015) Effektivnyi rynok truda kak element ekonomicheskoi bezopasnosti (The effective labour market as an element safety). Vestnik Kyrgyzsko-Rossiiskogo Slavyanskogo universiteta 15(8):145–148 (in Russ.)

Sidenko AG (2014) Obespechenie ekonomicheskoi bezopasnosti megapolisa instrumentami migratsionnoi politiki (na primere g. Sankt-Peterburga) (Provision of economic security of a megacity with the tools of migration policy (on the example of St. Petersburg)). Dissertation, Saint-Petersburg University of the Ministry of the Interior of the Russian Federation (in Russ.)

Solovieva NN (2016) Obespechenie ehffektivnoj zanyatosti trudovyh migrantov na rossijskom rynke truda (Ensuring effective employment of labour migrants in the Russian labour market). Dissertation, Tomsk State University (in Russ.)

Topilin A, Parfenceva O (2006) Migraciya i rynok truda v usloviyah krizisa (Migration and labour market in the crisis). Maks Press, Moscow (in Russ.)

Topilin AV (1975) Territorialnoye pereraspredelenie trudovykh resursov v SSSR (Territorial redistribution of human resources in the USSR). Ekonomika Publ, Moscow (in Russ.)

Tseykovets NV (2015) The level of the national economic security: index methods of the integral evaluation. Problems of modern economics 4(56):106–109

Tyuryukanova EV (2007) Rossiya budet sil'nee zaviset' ot truda migrantov (Russia will be more dependent on the labour of migrants). Otechestvennye zapiski 4(37). https://www.hse.ru/pubs/share/direct/document/74499710. Accessed 11 May 2017 (in Russ.)

Vakulenko ES, Leukhin RS (2015) Issledovanie sprosa na trud inostrannykh migrantov v rossiiskikh regionakhpo podannym zayavkam na kvoty (Investigation of demand for the foreign workforce in Russian regions using applications for quotas). Appl Econometrics 37 (1):67–86 (in Russ.)

Vakulenko ES, Leukhin RS (2016) Wage dicrimination against foreign workers in Russia. Ekonomicheskaya Politika 11(1):121–142

Vasil'eva AV (2015) Otsenka effektivnosti regulirovaniya mezhdunarodnoi trudovoi migratsii v regione (Evaluation of the effectiveness of international labour-migration regulation in the region). Reg Econ Theor Pract 38(413):20–32 (in Russ.)

Vitkovskaya G, Panarin S (eds) (2000) Migratsiya i bezopasnost' v Rossii (Migration and security in Russia). Interdialekt+, Moscow (in Russ.)

Vorobieva OD (2005) Politika i ehkonomicheskaya ehffektivnost' vneshnej trudovoj migracii (The policy and economic efficiency of external labour migration). In: Ispol'zovanie inostrannoj rabochej sily: problemy i perspektivy. IIEPS RAS, Moscow, pp 12–17 (in Russ.)

The Potential of the Kaliningrad Region in the Development of Health Tourism

Anna V. Belova⊙ and Irina V. Fedina-Zhurbina⊙

Abstract

The article studies the Kaliningrad Region from the perspective of health tourism development. The main purpose of the study is to explore the potential of the Kaliningrad region in the development of health tourism and its place in the Baltic sea region. In the course of the research, the recent publications on this issue were reviewed and the experience of both Russian and foreign researchers on the subject was analyzed; the cartographic method allowed to get clearer results, which are presented in the form of graphs and tables. The article considers natural and geographical preconditions for the development of health tourism in the Baltic sea region, analyzes resort facilities in the Kaliningrad region, shows the dynamics of attendance of the health resorts in the region from 2012 to 2016. The study shows the potential of each Kaliningrad municipality in the development of health and resort tourism. As a result of the study it was found out that the Kaliningrad region has a significant potential for the development of medical, health and health-related tourism in the Baltic sea region: a developed resort network, natural resources (mineral water, medicinal mud, etc.), advantageous geographical position, reasonable prices for resort and sanatorium-and-spa treatment). However, there is a need to expand the range of sanatorium-and-spa services of institutions in the Kaliningrad region to compete with similar institutions in the Baltic sea countries and neighboring Poland; also there is a need to improve the medical infrastructure and the level of medical care in the region, to expand the range of health and wellness services in the region.

Keywords

The Kaliningrad region · The Baltic region · Health tourism · Medical tourism · Resort facilities

A. V. Belova (✉) · I. V. Fedina-Zhurbina
Immanuel Kant Baltic Federal University, Kaliningrad, Russia
e-mail: ABelova@kantiana.ru

G. Fedorov et al. (eds.), *Baltic Region—The Region of Cooperation*,
Springer Proceedings in Earth and Environmental Sciences,
https://doi.org/10.1007/978-3-030-14519-4_32

Introduction

In recent years, health tourism in the world is gaining considerable intensity. More tourists tend to combine leisure activities and resort-and-spa treatment. Due to the favorable mild climate, availability of natural resources (mineral water, mud, peat) in the Baltic sea region, health tourism is actively developing (Khrupalo 2015). The most active countries in the development of sanatorium-and-spa and health tourism in the Baltic region on the basis of natural potential are Lithuania, Latvia, Estonia, Germany and Poland. Russia in the Baltic is represented mainly by the Kaliningrad region and the Leningrad region. Coastal resort complexes of these countries use the resources of both the Baltic Sea and other resources available in the region (mineral waters, medicinal muds, etc.). The Kaliningrad Region is one of the most favorable regions of Russia, developing health tourism in the Baltic region; it has a significant potential for sanatorium and resort development and, due to its economic and geographical location, is a promising and attractive region for the development of this type of tourism.

Theory

The notion of "medical" and "health" tourism is quite new, despite the fact that health services outside the places of permanent residence have been popular for quite some time. In research papers of both foreign and Russian authors you can meet many different interpretations of medical and health tourism. More precise definitions and delimitation of the concepts of health-related tourism can be found in the works of foreign authors. So, for example, abroad you can find the following concepts: "medical tourism", "wellness tourism", "health tourism", and interpretations of these concepts vary; having analysed the definitions given by foreign authors, we can conclude that the concept of health tourism is broader and includes the concepts of medical tourism and wellness tourism Hajioff (2007), Smith and Puczkó (2014). The World Tourism Organization (UNWTO) defines health tourism as a type of tourism, which includes services offered by SPA centers, or surgery-related medical services, associated with the most favorable price in comparison with the home region, and also the possibility to remain anonymous. According to the definition of Steve Haydzhof, "health tourism" is a trip to other countries or territories in order to obtain medical services and other healthcare activities, thereby he identifies this concept with the term "medical tourism". In contrast to Steve Haydzhof, the Greek author Ikos Aris defines health tourism as associated with visiting mineral and thermal springs for medicinal purposes. Two other definitions used by foreign authors, have a more specific meaning. Medical tourism is trips taken in order to improve the health or to treat the existing diseases, as well as economic activities which provide services that combine treatment and tourism (Bookman and Bookman 2007). As for the concept of wellness tourism, it was introduced in 1961 by Helbert Dann. This term refers to tourists getting health treatments, aimed at preserving and strengthening health associated with the use of natural resources, diet, and physical activity. For the Russian conceptual framework, this term most correlates, in our opinion, with the term "sanatorium-and-spa tourism", which prevailed in domestic works related to the study of medical and health tourism, and the concept of "medical tourism" became widespread only in recent years due to the active development of medical services offered to foreign citizens by various clinics, private medical and rehabilitation centers. Among Russian authors, giving the definition of tourism with the aim of improving health, one can mention Fedyakin (2001), who introduced the notion of "health tourism". One can also note such authors as Nabedrik (2005), Kazakov (2002), Sukhov (2002), Gulyaev (2011), Navodnichiy (2011) and others, who considered theoretical approaches to the concept of medical and health tourism. Among the scientists studying the development of medical and health tourism in Russia, one can distinguish Sokolova et al. (2017) and others. Issues of sanatorium and resort tourism are discussed in the works of such Russian authors as Arkhipov and Sevryukov (2013), Mozokina (2012), Nikitina, and foreign authors—Smith and Kelly (2006), Erfurt-Cooper and Cooper (2009), Voigt and Pforr (2013), etc. The combination of the development of health and sustainable tourism is considered in the works by Rassokhina and Seselkin (2016) and Kropinova (2016). Sanatorium and resort tourism and its development in the Kaliningrad region and in the Baltic region are explored in their works by Kropinova (2008), Mitrofanova (2010), Nikitina (2012), Oborin and Mingaleva (2017) and others.

Research Methods

The given research has used the statistical analysis methods, conducting and processing the interview (in comparing the official statistics and determining the actual results for the allocation of functioning facilities for the sanatorium complex in the Kaliningrad Region). The authors interviewed more than 30 representatives of sanatorium and resort institutions of the region were conducted and processed. When processing statistical material, the method of graphical data visualization (drawings, graphs) was applied. In addition, a cartographic method was used to visualize the economic and geographical position of the Kaliningrad region in the Baltic sea region.

Results

The Baltic sea region with its mild and cool climate, has long been the optimal destination for developing medical and recreational tourism with the focus on sanatorium and resort vacation. In addition to the Baltic Sea as a resource, the sanatorium-and-spa resorts of the Baltic region also use health-improving resources like mineral waters and medicinal muds, the reserves of which are available in the region in sufficient quantities (Kornevets et al. 2008). Countries of the Baltic sea region with the most developed sanatorium and spa tourism are Germany, Poland, Lithuania, Latvia and Estonia. In Lithuania, Latvia and Estonia, sanatorium and spa tourism is integrated into the Baltic sanatorium and resort health tourism cluster (Smith 2015).

The Kaliningrad region is one of the regions of Russia in the Baltic Sea region, which has a high potential for the medical tourism development, with a historically developed diverse sanatorium complex and an advantageous economic and geographical location (Fig. 1).

In their properties, the resorts of the Kaliningrad region are not inferior to the neighboring Lithuanian ones. As early as the beginning of the 19th century, the resorts of Krantz (now Zelenogradsk) and Rauschen (now Svetlogorsk) were the main destinations in East Prussia, and very popular with the Germans. After 1945, Zelenogradsk and Svetlogorsk were converted into federal resorts (according to the Decree of the Government of the Russian Federation of March 29, 1999,

Fig. 1 Position of the Kaliningrad region in the Baltic Sea Region. Resorts of the Kaliningrad region. *Source* Compiled by the author

No. 359 "On the recognition of the health resorts of Zelenogradsk and Svetlogorsk-Otradnoye, located in the Kaliningrad region, as the resorts of federal significance.").

In the Kaliningrad region, there are 20 operating facilities for sanatorium and health resorts. Almost all of them are located in the coastal cities and settlements of the region. These are sanatoriums and hotels with spa and wellness centers. The main objects of the sanatorium-and-spa infrastructure, as well as their specialization are presented in Table 1.

From the table presented, it is clear that the Kaliningrad region has 20 functioning institutions that provide health services with treatment, including diagnostic procedures. There are only 4 health institutions specializing on children's treatment, but other sanatoria let the parents stay with their children. The total capacity of all operating sanatorium-and-spa treatment facilities in the Kaliningrad region is 3524 people. Sanatoriums work year-round, which is a strong point of development of medical tourism in the region. At the same time, several sanatorium-and-spa institutions are currently not functioning: children's sanatorium "Maisky" (150 people capacity) and children's cardio-rheumatology sanatorium (60 people capacity).

It should be noted that for the last 5 years the peak of attendance of sanatoriums and resorts in the Kaliningrad region falls on 2015 (Fig. 2). The attendance of Russian citizens has increased, and that of foreign citizens—declined (Fig. 3).

As can be seen from Fig. 2, since 2016 the number of people receiving sanatorium-and-spa treatment has decreased. First of all, this is due to a reduction in the costs of enterprises to provide employees with vouchers to sanatoriums and health resorts, as well as a decrease in people's incomes as a result of the economic crisis.

The decrease in the number of foreign citizens who visited the sanatorium and resort facilities of the Kaliningrad region is associated with the economic crisis and the interruption of a simplified visa regime (cancellation of 72-h visas and local border traffic with Poland).

In the Kaliningrad region, possessing a number of advantages and resources for the development of sanatorium-and-spa tourism, these institutions are located only in coastal cities and settlements. However, other municipalities located to the east of Kaliningrad have a number of prerequisites for building quality accommodation facilities with a curative and health-improving component (Ikkos 2002; Kelly 2010). Table 2 presents the potential for the development of health tourism in the municipalities of the Kaliningrad region (excluding coastal municipalities).

The presented table shows that the region has a significant potential and the prerequisites for the development of health-improving and spa-resort tourism. Moreover, other advantages include the compactness of the region and the developed transport network.

However, a number of existing sanatoriums on the coast of the region are in a deplorable state, they require repair and renovation. Thus, it is expedient to develop the existing sanatorium and resort complex in the coastal municipalities of the region, increasing the region's competitiveness in health tourism in the Baltic sea region.

Table 1 Infrastructure of the sanatorium and resort complex of the Kaliningrad region and its specialization

No.	Name of institution	Specialization	Capacity (percentage of the total number of accommodation facilities in the Kaliningrad region)
Zelenogradsk			
1	Sanatorium "Zelenogradsk"	– Cardiovascular system – Nervous system – Musculoskeletal system – Gynecological diseases	235 people (1.8%)
2	Sanatorium "Chaika"	– Cardiovascular system – Nervous system – Musculoskeletal system – Gynecological diseases	190 people (1.4%)
3	Sanatorium "Teremok"	For children – Central and peripheral nervous system with speech, developmental disorders, motor disorders, cerebral palsy. (psychoneurological profile)	100 people (0.8%)
Svetlogorsk			
4	Sanatorium "Yantarny Bereg"	– Cardiovascular system – Musculoskeletal system – Nervous system – Gastrointestinal tract – The respiratory system – ENT-diseases – Gynecological diseases – Metabolic disorders – Endocrine system	650 people (2.4%)
5	Sanatorium "Yantar"	– Nervous system – Musculoskeletal system – Cardiovascular system – Gynecological diseases – The respiratory system – Urology – Gastrointestinal tract	300 people (2.3%)
6	Svetlogorsk Central Military Sanatorium	– The circulatory system – Musculoskeletal system – Nervous system – Gynecological diseases – diseases of the digestive system	320 people (2.4%)
7	Sanatorium "Svetlogorsk"	– Cardiovascular system – Musculoskeletal system – Nervous system and diabetes mellitus	146 people (1.1%)

(continued)

Table 1 (continued)

No.	Name of institution	Specialization	Capacity (percentage of the total number of accommodation facilities in the Kaliningrad region)
8	Pension «Baltika»	– Nervous system – Gastrointestinal tract – Musculoskeletal system	160 people (1.2%)
9	Pension "Volna"	– Diseases of the respiratory system – Nervous system – Musculoskeletal y system – Circulatory system	200 people (1.5%)
10	Hotel "Olimp"	– Educational and health-improving complex – Musculoskeletal system – Respiratory system	60 people (0.5%)
11	Pension "Rauschen"	– Cardiovascular system – Musculoskeletal system – Digestion and respiratory systems	60 people (0.5%)
12	Sanatorium "Trojka"	– Diseases of the musculoskeletal system – The respiratory system – Nervous system	60 people (0.5%)
13	Sanatorium-dispensary "Energetik"	– Musculoskeletal system – Cardiovascular system – Nervous system – Gastrointestinal tract	65 people (0.5%)
15	Children's tuberculosis sanatorium	– Tubercular pathology and its complications	60 people (0.5%)
16	Anti-tuberculosis sanatorium of the Kaliningrad region	– Pulmonary forms of tuberculosis	60 people (0.5%)
Otradnoe settlement			
17	Social and Wellness Center "Mechta"	– Cardiovascular system – Nervous system – Musculoskeletal system	100 people (0.8%)
18	Sanatorium "Otradnoe"	– Cardiovascular system – Organs of movement – Nervous system – Digestive system – Endocrine system	300 people (2.3%)
19	Children's pulmonological sanatorium "Otradnoe"	– Diseases of the respiratory system	150 people (1.12%)
Pionerskiy			
20	Pediatric orthopedic sanatorium "Pionersk"	– Musculoskeletal system	300 people (2.3%)

Source Compiled by the author

Fig. 2 The number of tourists who visited health resorts of the Kaliningrad region in 2012–2016. *Source* Compiled by the author on the basis of data from the Federal State Statistics Body

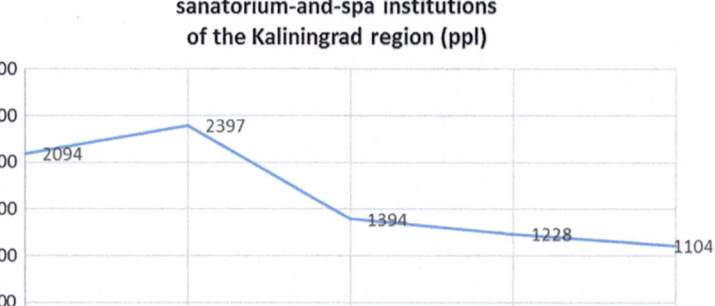

Fig. 3 Number of foreign citizens who visited sanatorium-and-spa institutions of the Kaliningrad region. *Source* Compiled by the author on the basis of data from the Federal State Statistics Body

Conclusion

As the study shows, the Kaliningrad region is one of the regions of Russia that has significant potential for the development of medical and health tourism. Basically, this is a type of health tourism based on the use of climatic and balneological resources (the Baltic Sea, mineral waters, medicinal muds). In the region, health tourism has been developing for a long time, and it is a historically established destination for receiving sanatorium-and-spa services both for Russian tourists (initially) and for foreign tourists (beginning in 2000).

Table 2 Resources and potential for the development of sanatorium and health tourism in the Kaliningrad region (excluding coastal cities)

No.	Municipality	Resources for sanatorium development	Proposals for the development
1	Slavsky	– Mineral water	– Baths, mud baths, spa centre
2	Sovietsky	– The Neman river – A border town with a border-crossing checkpoint (Lithuania)	– Rehabilitation center for children with disabilities, development of a similar center for adults
3	Chernyakhovsky	– three rivers (Angrapa, Instruac, Pregola) – Accessibility	– A spa center, a rehabilitation center for adults
4	Mamonovsky	– Accessibility – Border town (Poland) – Vistula lagoon	– A spa complex on the coast of the Vistula lagoon
5	Nemansky	– Mineral water aquifer for balneological use	– A sanatorium complex for children based on the "Sputnik" health camp
6	Nesterovsky	– Rominten forest – Vishtynetskoye lake, the river Krasnaya, the river Pissa	– A spa center, a wellness center
7	Ozersky	– Natural resources (forests, landscapes)	– Development of the winter sports resort and a rehabilitation center
8	Polessky	– The Couronian Lagoon – Medicinal muds	– Health resort using the medicinal muds – Mud baths
9	Pravdinskiy	– Border town (Poland) – The river Lava (river Lyna in Poland)	– A spa center specializing in heart diseases
10	Svetlovsky	– Kaliningrad (Vistula lagoon) – Forest resources	– The creation of a children's center for the rehabilitation of older children and adolescents
11	Bagrationovsky	– Natural resources (lakes) – Proximity to the border – Accessibility	– The development of the center for the prevention of respiratory diseases
12	Gvardejsky	– Natural resources (the river Deima) – Accessibility	– Rehabilitation sanatorium for children with respiratory system problems
13	Guryevsky	– The Pregola river – Accessibility (part of the Kaliningrad agglomeration)	– Post-traumatic rehabilitation center for the elderly
14	Gusevsky	– Underground mineral water – The proximity of the border with Poland	– Balneary
15	Krasnoznamensky	– Ecologically clean forest area	– A sanatorium with a specialization in the treatment of diseases of the respiratory system

(continued)

Table 2 (continued)

No.	Municipality	Resources for sanatorium development	Proposals for the development
		– The Sheshupe and Neman rivers – The State Monument of Nature "Dobrovolsky"	
16	Ladushkinsky	– Vistula lagoon – Accessibility	– Creation of a health-improving year-round center based on the "Chaika" health camp

The coastal areas of the region are most developed in terms of infrastructure, federal resorts and almost all sanatoriums of the Kaliningrad region (sanatoriums, hotels with sanatorium services) are located there. In the total accommodation capacity (number of beds), sanatorium-and-spa institutions occupy …%. This indicator, in our opinion, is sufficient for the region, closing the needs of tourists in sanatorium-and-spa services. However, a significant part of the infrastructure needs complete reconstruction, some accommodation facilities need to be converted to meet the needs of people with disabilities. Another necessary measure to make the sanatorium-and-spa complex of the region more attractive for tourists is the change of concepts for the existing institutions. For example, accommodation facilities in resort cities lack loyalty programmes and means for attracting tourists, such as ready-made packages of services aimed at different categories of tourists. A good example of such attraction of tourists can be the sanatorium and spa complexes of neighboring Lithuania (in Palanga, Druskininkai and other cities).

Since the Kaliningrad region possesses not only coastal resources, but also a number of natural resources that can be used for therapeutic purposes, it is advisable to develop health-improving and spa centers in the central and eastern part of the region, and to attract investments for the formation of a health infrastructure on the basis of mud therapy for rehabilitation procedures. Taking into account the compactness of the region and the developed transport accessibility, treatment and rehabilitation in potential centers in the east of the region can be combined with trips to the coast.

Today, sanatorium-and-spa tourism on the coast of the Baltic Sea is the most developed kind of medical and health-improving tourism in the Kaliningrad region. Other types of medical tourism, aimed at foreign citizens, can be developed in the region due to its favorable economic and geographical situation; but the regional medicine is not so much advanced, although there are prerequisites for its development. To improve the image of the region as an attractive center for health and resort tourism, first and foremost, it is necessary to bring the sanatorium and resort infrastructure and services to a high level, which would be comparable with the level of similar institutions in neighboring Lithuania.

References

Arkhipov AE, Sevryukov IYu (2013) Marketingovye aspekty formirovaniya mnogokonturnogo sanatorno-kurortnogo produkta (Marketing aspects of the formation of a multi-contour sanatorium-spa product). Problemy sovremennoi ekonomiki 3(47):269–272 (in Russ.)

Bookman M, Bookman K (2007) Medical tourism in developing Countries. Palgrave MacMillan, New York

Erfurt-Cooper P, Cooper M (2009) Health and wellness tourism: spas and hot springs. Channel View Publications

Fedyakin AA (2001) Teoretiko-metodicheskie osnovy ozdorovitel'nogo turizma (Teoretiko-methodical bases of improving tourism). Dissertation, Maikop

Gulyaev VG (2011) Strategiya razvitiya turizma i turistskih destinacij–effektivnyj instrument antikrizisnogo upravleniya. Vestnik RMAT 2(2)

Hajioff S (2007) Health tourism: 100 years of living. http://www1.imperial.ac.uk/resources/4042628D-B846-4531-BB7D0F8293266C76. Accessed 15 Dec 2017

Ikkos A (2002) Health tourism: new challenge in tourism. http://www.gbrconsulting.gr/articles/Health%20Tourism%20-%20a%20new%20challenge.pdf. Accessed 9 Apr 2018

Kazakov VF (2002) Opyt organizatsii lechebnogo turizma v sanatorii srednei polosy Rossii (Experience in the organization of medical tourism in the sanatorium of the central Russian region). Kurortnye vedomosti 2:10–12 (in Russ.)

Kelly C (2010) Analysing wellness tourism provision: a retreat operators' study. J Hospitality Tourism Manage 17(1):108–116

Khrupalo VM (2015) Problemy i perspektivy razvitiya meditsinskogo turizma v Kaliningradskoi oblasti (Problems and prospects of development of medical tourism in the Kaliningrad region). Nauchnye issledovaniya: ot teorii k praktike 2(4):230–233 (in Russ.)

Kornevets VS, Kropinova EG, Dragileva II (2008) Turistskoe raionirovanie Kaliningradskoi oblasti (Tourist district of the Kaliningrad region). Geografiya i turizm 1:117–132 (in Russ.)

Kropinova EG (2008) Vestnik Rossijskogo gosudarstvennogo universiteta im. Immanuila Kant (7):85–88

Kropinova EG (2016) Transgranichnye turistsko-rekreacionnye regiony na Baltike [Cross-border tourist and recreational regions in the Baltic sea]. Kaliningrad, p. 272

Mitrofanova AV (2010) Regional'nyi turistskii klaster kak forma prostranstvennoi organizatsii turizma (na primere Kaliningradskoi oblasti) (Regional tourist cluster as a form of spatial organization of tourism (on the example of the Kaliningrad region). http://www.kantiana.ru/postgraduate/announce/avt_mitrofanova.pdf. Accessed 9 Apr 2018 (in Russ.)

Mozokina SL (2012) Perspektivy razvitiya ozdorovitel'nogo turizma (Prospects for the development of health tourism). Izvestiya Sankt-Peterburgskogo gosudarstvennogo ekonomicheskogo universiteta 2:89–93 (in Russ.)

Nabedrik VA (2005) Geografiya lechebnogo turizma v Evrope: modeli razvitiya i transformatsionnye protsessy (Geography of health tourism in Europe: development models and transformation processes). Dissertation, Moscow (in Russ.)

Navodnichiy RM (2011) Sistemnyi analiz ponyatiya « lechebno-ozdorovitel'nyi turizm » (System analysis of the concept of "medical and health tourism"). Vestnik universiteta (GUU) 1:142–147 (in Russ.)

Nikitina O (2012) Innovative trajectory of formation investment projects of tourist cluster in Russia. In: Scientific enquiry in the contemporary world: theoretical basics and innovative approach. pp 124–127

Oborin MS, Mingaleva ZhA (2017) Napravleniya i tendentsii razvitiya sanatorno-kurortnogo kompleksa i lechebno-ozdorovitel'nogo turizma yuzhnykh regionov Rossii v nestabil'nykh sotsial'no-ekonomicheskikh usloviyakh (Directions and tendencies of the development of the sanatorium and health resort tourism in the southern regions of Russia in unstable social and economic conditions). Nauchnye vedomosti Belgorodskogo gosudarstvennogo universiteta. Seriya: Ekonomika. Informatika 41(2):32–37 (in Russ.)

Official website of the Government of the Kaliningrad region. https://gov39.ru. Accessed 12 Dec 2017 (in Russ.)

Rassokhina TV, Seselkin AI (2016) Analiz sovremennykh problem i prioritetov v oblasti ustoichivogo razvitiya turizma (Analysis of current problems and priorities in the field of sustainable tourism development). Ekonomika ustoichivogo razvitiya 1(25):318–323 (in Russ.)

Smith M, Kelly C (2006) Wellness tourism. Tourism Recreation Res 31(1):1–4

Smith M, Puczkó L (2014) Health, tourism and hospitality: spas, wellness and medical travel. Routledge

Smith MK (2015) Issues in cultural tourism studies. Routledge

Sokolov AS, Man'ko NP, Gulyaev VG (2017) Teoretiko-metodologicheskie aspekty meditsin- skogo turizma (Theoretical and methodological aspects of medical tourism). Vestnik RMAT 3:105–111 (in Russ.)

Sukhov RI (2002) Osobennosti razvitiya i sovremennoe sostoyanie turizma v Rostovskoi oblasti (Features of development and the current state of tourism in the Rostov region). Dissertation abstract, Rostov on Don

Territorial body of the Federal State Statistics Service for the Kaliningrad Region. http:// kaliningrad.gks.ru/wps/wcm/connect/rosstat_ts/kaliningrad/ru/statistics/sphere/. Accessed 12 Dec 2017 (in Russ.)

Voigt C, Pforr C (eds) (2013) Wellness tourism: A destination perspective, vol 33. Routledge

Printed by Printforce, the Netherlands